# Energy and Analytics

## BIG DATA and
## Building Technology Integration

# Energy and Analytics

## BIG DATA and
## Building Technology Integration

John J. "Jack" Mc Gowan, CEM

**THE FAIRMONT PRESS, INC.**

CRC Press
Taylor & Francis Group

# Library of Congress Cataloging-in-Publication Data

McGowan, John J., 1950-
  Energy and analytics : BIG DATA and building technology integration / John J. "Jack" McGowan, CEM.
        pages cm
  Includes bibliographical references and index.
  ISBN 0-88173-744-5 (alk. paper) -- ISBN 0-88173-745-3 (electronic) -- ISBN 978-1-4987-4429-4 (Taylor & Francis distribtion : alk. paper) 1. Buildings--Energy conservation. 2. Energy auditing--Data processing. 3. Buildings--Energy consumption--Data processing. 4. Big data. I. Title.

  TJ163.5.B84M395 2015
  696.0285'57--dc23

                                                                    2015015301

Published by The Fairmont Press, Inc.
700 Indian Trail
Lilburn, GA 30047
tel: 770-925-9388; fax: 770-381-9865
http://www.fairmontpress.com

Distributed by Taylor & Francis Ltd.
6000 Broken Sound Parkway NW, Suite 300
Boca Raton, FL 33487, USA
E-mail: orders@crcpress.com

Distributed by Taylor & Francis Ltd.
23-25 Blades Court
Deodar Road
London SW15 2NU, UK
E-mail: uk.tandf@thomsonpublishingservices.co.uk

Printed in the United States of America
10 9 8 7 6 5 4 3 2 1

ISBN 0-88173-744-5 (The Fairmont Press, Inc.)
ISBN 978-1-4987-4429-4 (Taylor & Francis Ltd.)

# Dedicated

to
Judy
...with the love of a lifetime
and gratitude for your support in all things.

# Table of Contents

# Author's Note on How this Book is Organized

Given the complex nature of the energy and analytics topic, and the content provided here, it is appropriate to briefly discuss the format and use of this book. Energy management in the 21st century is being influenced by wide ranging policy, business and technology trends; none more significant than analytics and big data. Building owners, operators and Certified Energy Managers (CEM) can easily be daunted by the complexity of analytics implementations, as they entail a number of steps to fully leverage the opportunity. An important factor is that energy and analytics technology cannot produce optimum results, unless it can access an underlying fabric of building/energy technology, and that technology must be fully interoperable and operational. This simple fact explains a great deal about the organization of this book.

This book is organized in three sections. Section I is an introduction to and thorough treatment of the energy and analytics market opportunity. Chapters 1 through 7 define analytics and related technology available in the market today, as well as how they may be leveraged for optimum economic value. Equally important, it articulates the energy context, which primarily involves electricity. Electricity is the highest quality form of energy because it offers the cleanest and most diverse use. It is essential to business, education and entertainment, and the buildings in which all of these industries operate. This is because electricity is the primary energy source for the information economy, powering smart phones, tablets, computers, the "Cloud" and everything in between. In other words, electricity is central to the 21st century lifestyle, and Section I also describes changes in the way electricity is purchased and sold, which demands analytics technology. An important note for this section, and those that follow, is that every effort was made to arrange content in the best order possible, so that more general topics are covered prior to more complex. The idea is to start with a high-level view of the market, industry and technology, and then progress to discussions of how this technology may be used for critical applications.

Section II of the book focuses on the fabric of underlying building and energy technology that is critical for analytics to drive maximum value. Because these topics are both interdependent with, and essential

xiii

to, energy and analytics, they warrant an introduction from one of the most influential people in building systems, Ken Sinclair, founder of www.automatedbuildings.com. His knowledge of building automation systems (BAS) and energy engineering is unparalleled, making him uniquely qualified to introduce Chapters 8 through 18. He can also provide insight on how these technologies are interconnected and interoperable, and what that can mean for energy and analytics. These underlying systems must exist to capture enough data for meaningful analytics. In many cases they are also required to execute the changes in building operations and energy consumption recommended by analytics engines. Chapters in this section discuss the full range of technologies that are critical to effective implementation of energy and analytics, as well as related topics including the Internet of Things (IoT), financing for energy projects, and measurement and verification.

Section III of the book offers case studies in energy and analytics, to provide a glimpse of what success looks like. James M. Lee, CEO of Cimetrics, Inc., has decades of real world experience with analytics, as it is discussed here. No one is better able, than Mr. Lee to write an introduction to the topic of analytics success. These chapters present readers with real examples of how this technology is producing dramatic results in a wide range of building and campus types. Mr. Lee has perhaps the longest and most technically diverse tenure in this industry, while also being one of its most successful business leaders. Under Mr. Lee's leadership Cimetrics, Inc., has deployed, and been an industry leader, in energy and analytics solutions. These projects span a wide range of applications including: large campuses with a who's who of higher education, public and private hospitals, industrial/pharmaceutical facilities, and a wide range of buildings. The stories in these chapters also come from all types of buildings and campuses, and show what is possible with energy and analytics. These are but a glimpse of the operations and business value of this technology, and the breadth of applications to which it's been applied.


PROFESSIONAL ACKNOWLEDGEMENTS

Over nearly four decades working with energy, buildings and technology is has been my sincere honor to meet and work with thought leaders, visionaries and exceptional professionals from across multiple industries.

There are a number of these people without whom this book could not have been written. Particular thanks are due to:

**John Petze** for writing the Book Foreword, and for making contributions to my work based on his knowledge and business acumen in the form of; written material, edits and an unparalleled industry perspective. **Jim Lee** for writing the Case Study Foreword and for his amazing industry insights, as well as his contribution of written material, and for sharing his incomparable perspective on the far reaching impacts that this technology is having on many industries. Thanks also to Jim's team at Cimetrics including **Mary Ellen Cantabene, Karim Bibawi** and **David Landman** for their contributions. **Ken Sinclair** for writing the Technology Section Foreword and for making the most significant advancement to the automation industry in decades, through creation of www.automatedbuildings.com. This book drew from his writings on automatedbuildings, as well as from articles published there by Anno Scholten, Jim Sinopoli and Nirosha Munasinghe. **Anno Scholten,** in particular is a true visionary and serial entrepreneur in the building intelligence and energy space. Ken Sinclair has attracted a host of gifted writers such as those above to www.automatedbuildings.com. His vision in starting an Internet-based magazine, industry portal and worldwide gathering place nearly twenty years ago is a marvel. **Glen Allmendinger** for sharing a wealth of business and technology knowledge in his excellent chapter on the Internet of Things. **Leighton Wolffe** for allowing the ESCO 2.0 work to be incorporated with the finance chapter, and to Leighton and **Peter Kelly-Detwiler** for sharing their vision for the future of electricity that influenced the chapter on that topic. **Professor Jim Lee** and his team at the University of Louisiana at Lafayette for the Six Sigma chapter. **Kurt Yeager** for introducing me to **Bob Galvin** and for giving me the opportunity to work with them for three years. Their passion for excellence, and for the transformation of the electricity business, provided major inspiration for this book. Their selfless sharing of that vision for the future was critical to forming this view of a 21st Century Electric System. **Tom Shircliff** for sharing what the future for Smart Cities can be, in his chapter on Envision Charlotte. **Darell Smith** for taking that Smart City view to the next level, by showing how corporations and cities could join forces to change the world, in the chapter on Microsoft 88 acres. **Ron Rajecki and Contracting business** for inclusion of the Smart Grid Smart People story that is so illustrative of success for these ideas and technologies. **Pat**

**Gibson and Luanne Garcia,** partners and friends who made it possible for me to pursue excellence, and create projects that exemplified my vision for the future, in the form of projects, business and technology. **Dave Flood, John Christensen, Andy Watson, Bobbie Gutierrez, Leslie Carpenter, Lisa Randall, Don Swick and Andrea Mammoli, PhD** are also important to thank as visionary executives and building managers, who embraced the leading edge in energy management and sustainability, and who I had the great honor to serve with exciting projects. **Brian Goodchild, John Nicholls, Scot Stickle, Jay Garbarino and Drew Perrin** at Delta Controls Inc. and Coppertree Analytics for sharing content and industry perspectives selflessly to improve this book. **Nick Michaels** for his vision, courage and singular business skill. **Jim Young and Howard Berger** for creating the premier industry venue for building intelligence, in Realcomm and IBcon, allowing professionals to share insight and success. **Anto Budiarjo** for vision and fortitude in setting the bar for what a truly excellent industry event should be, and for continuing that work with www.PointView.com. **Bob Graves** of the Governing Institute for inviting me to speak and participate with them in Energy Roundtables around the country, that made it possible to interact with cities pursuing excellence in energy initiatives. **Julia Burrows,** Director of the Governing Institute for continuing to include me with Big Data programs and as an advisor on the Institutes work in the pursuit of infrastructure renewal. **Jim Heath** for his vision and exceptional business and finance acumen in creating a solar business model like no other. **Santa Fe Mayor Javier Gonzales** for his vision, resolve and leadership in creating a ground breaking initiative, called the Climate Action Task Force, to create an unparalleled vision for the future, and for giving me the incomparable opportunity to work on energy and sustainability at the community level. The Mayor-appointed Co-Chairs; **City Councilor Peter Ives and Former Mayor David Coss,** who working together with **Director of Utilities Nick Schiavo and Renewable Energy Planner John Alejandro** assembled an unparalleled team including: **Randy Grissom, Beth Beloff, David Van Winkle, Kathleen Holian, Glenn Schiffbauer, Rob Hirsch, Regina Wheeler, David Breeker, Harold Trujillo, Craig O'Hare and Linda Smith. Eric Lightner,** visionary leader of the U.S. Department of Energy (DOE) Smart Grid task force. **Steve Widergren, Rik Drummond, Lynne Kiesling, Tracy Markie, Wade Troxell, Alison Silverstein, Richard Schomberg, Nora Mead Brownell, Erich Gunther, Dave Cohen, Ron Ambrosio, Ward Camp, Tracy Markie** for their leadership and contributions as members of the DOE GridWise Architecture Council. **Dave Frederick** for his dedication and taking on the leadership of Energy Control. **Bruce Cantrell** for founding a business in ECI that was ideally positioned for growth, and infusion of new leadership, but never lost sight of the customer. **Rob Masinter** for his tenacity in pursuing a deal and his insatiable appetite to understand businesses and markets. **Barney Capehart** for graciously allowing me to use John Petze's chapter from his book. **Mahesh Ramanujam** and **Brendan Owens** at USGBC for the vision to see the next frontier for green buildings in PEER and acquiring **John Kelly** and his team from the Galvin Electricity Initiative to lead the effort. **Marc Petock** for being the distinguished gentleman of building technology and for generously sharing his knowledge of the industry. **Terry Swope** and his team at LynxSpring including **Bob Mealey**. **Greg Dixon** of EnerNOC for sharing industry insight and allowing me to speak and participate in EnergySmart events. **B. Scott Muench** for his amazing technical prowess and having the unique combination of tech smarts with sales and marketing skill.

# Part I— Foreword

*By John Petze*

There can be no turning back. The age of data has arrived and not a moment too soon for the buildings that provide the environment for our modern lives.

For decades now we have witnessed the development of building systems that attempt to exploit the power of the microprocessor to deliver better performance. Analog controls became digital, sensors became more precise, controllers became networked and connected to computer screens so we could "see" what our systems were doing—in 32 million colors no less. Each technological advance promised the potential to improve the operation of our facilities and reduce energy costs, yet the consistency of those results has never met our expectations.

Throughout this digital evolution, however, I think it is fair to say that few of us saw the impact that the data produced by these systems would have. That is changing. The growing realization of the value of the data produced by building equipment systems is creating growing interest from owners and operators, along with the hype and confusion attendant to all new technologies. Its especially fitting then that one of the most respected energy engineers of our times is now tackling the topic to bring clarity and understanding to the people responsible for designing, constructing, operating and maintaining the facilities we live, work, learn and play in.

The reality is that the people responsible for managing buildings rarely have the time to be expert in all of the systems and technologies that populate their buildings. And they shouldn't have to. Instead, they need to be able to rely on trusted experts to guide and advise them and insure that all of this promising technology operates effectively. Those experts in turn have a responsibility to use good engineering principles to insure their recommendations and counsel are correct. Fundamental to achieving that goal is the need to use data to continuously validate and verify the results of their designs, control strategies and recommendations. In short, we have an obligation to use data effectively. While it could be argued that this obligation isn't new, what is new is that technology has given us the tools to do that effectively and consistently.

The explosion of data available from the myriad "smart devices" found in today's buildings demands a convergence—bringing together, or perhaps forcing together, professionals with different knowledge, training, experiences and responsibilities—a shotgun marriage of IT, mechanical, control, and financial disciplines. Only through the combined efforts of all parties involved in the process of designing, building and operating buildings can we achieve the efficiency and sustainability that investors, operators, occupants and society deserve. It is a fundamental responsibility that requires we effectively use this new data resource.

Buildings are complex. No one knows it all—although Jack does come close. Enjoy the read. You will profit greatly from it.

*John Petze, C.E.M., is a partner at SkyFoundry, the developers of SkySpark®, analytics for building, energy and equipment data. John has over 30 years of experience in building automation, energy management and the Internet of Things, having served in senior positions for manufacturers of hardware and software products including: President & CEO of Tridium, VP Product Development for Andover Controls, and Global Director of Sales for Cisco Systems Smart and Connected Buildings group. At SkyFoundry he helps owners take advantage of advanced operational analytics to create truly intelligent buildings.*

# Chapter 1

# The 21st Century Energy Marketplace and Facilities

The energy marketplace has seen dramatic changes over the last three decades. A combination of market and business developments have occurred, but these have been complimented by explosive growth in technology. Information technology (IT) has been at the center of many of these changes, and in particular web services, but those developments are only part of the story. Focal points for this book will be around energy, particularly electricity, and the impact that the deregulation of electricity had on nearly half of the United States, and on the internet with web enablement of nearly every piece of building equipment, especially with the internet of things. Completely intertwined with that discussion of web services and the internet for the world wide web, is coverage of data communication standards for all of the equipment inside the building, and the opportunities that these developments present for building managers as well as business managers. Standards covered here, in particular BACnet for BAS and web services for nearly everything, have made it possible to both communicate with, and extract data from, any piece of equipment or building. The focus here is on deriving value from this data-rich environment through analytics, which is the latest technology development taking advantage of IT and the internet for energy management. The term "big data" has received huge attention across many domains in recent years. Most consumers seem to understand the notion of using information to track buying preferences, similar to the way companies like Amazon.com™ do, and then using those data to enhance the buying process. Most people have had the experience of searching for a book online, and then being contacted later with a recommendation for a book they might like. With energy, a similar set of opportunities exists to improve value, but analytics is much more than a trend, it is actually a crescendo of technological development that has unified systems throughout the energy and buildings landscape, with

3

particular emphasize on electricity. Recovery from the great economic downturn that began in 2008 was slow and arduous. This affected building managers trying to get approval for capital renewal, and it affected businesses alike. Analytics has been a great opportunity to bring value in an environment where managers wanted to do more with less, and it appealed to both groups. The idea of managing and measuring buildings to a set of key performance indicators (KPIs) has also been very appealing to building owners and operators.

Analytics brought value to owners in many ways, but in general there have been four predominant applications for the technology:

- Energy management,
- Fault detection and diagnostics (FD&D),
- Regulatory compliance, and
- Meeting organizational objectives

This book focuses primarily on the energy application for analytics, but that topic closely aligns with FD&D because equipment, whether mechanical or electrical, accounts for the vast majority of energy consumption in buildings.

Energy management is a primary use for analytics technology, and will be covered in great detail throughout the book, beginning with the next chapter. That chapter, entitled 4$^{Plex}$ Energy™, begins to define the complexity that is found in the 21$^{st}$ century energy marketplace, and the value that analytics technology can offer. This topic is the next generation of enterprise energy management, and is being integrated into a wide range of energy services. The goal of this book is to open the reader's eyes to the significant opportunity that this technology can offer.

Analytics for fault diagnostics and detection (FD&D) may be the first topic readers think of when energy and analytics is mentioned. FD&D is an excellent application for analytics, especially relating to heating ventilation and air conditioning (HVAC), because of the complexity associated with equipment control and operation, as well as the environmental control of buildings for comfort, indoor air quality and cost.

Analytics for regulatory compliance is another related topic, particularly with FD&D. Perhaps the best example of this would be joint commission compliance requirements for hospitals. Joint commission evaluates hospitals for accreditation and certification based upon meeting certain performance standards. There are numerous examples of buildings that must meet a performance standard to be licensed, to meet health or other codes, and for many other reasons. Analytics is a highly effective way to establish performance and to document it over time, when regulatory compliance is required. In many respects this process is similar to energy or FD&D, and in many cases compliance means operating equipment to meet certain environmental or other conditions. In that way, using analytics for regulatory compliance is not unlike the above. However, since the focus here is on energy, details relating to the use of analytics to meet compliance are not be covered in this book.

Meeting organizational objectives and key performance indicators (KPIs) is the last common use of analytics. The objectives may relate to measurement of certain aspects of a facility such as sustainability. Tracking carbon emissions, etc. can be done very easily using analytics. This is because energy consumption units for customer facilities are already in the database, so it is a simple calculation to assess carbon. This creates real synergy between energy analytics and sustainability. Another example of organizational objective is vendor performance. Many managers have realized that it is possible to develop a vendor oversight program using analytics. Consider the example of an HVAC service contractor that the owner wanted to evaluate. Measurable data are already being captured by analytics regarding environmental quality, equipment runtime, energy consumption or repair intervals. It is possible to develop rules that can be evaluated based on operational data and very quickly indicate how well the vendor is performing. There are any number of similar organizational objectives that may lend themselves to analytics. Again, since the focus here is on energy, details relating to the use of analytics to meet compliance will not be covered in this book.

Dashboards are a final related topic to energy and analytics. A dashboard of some type is typically integrated with energy and analytics technology, and these will be discussed further throughout the book.

To simplify discussion of the complex array of systems deployed as analytics systems, a blanket term, *"energy and analytics,"* will be used here. Energy and analytics will be the lens through which technology and operational discussions unfold throughout the book. Electricity is the highest quality form of energy, which makes it most essential for a 21st century quality of life, including technology for education, business and entertainment. At the heart of this discussion is a new marketplace that reflects human lifestyle changes, but equally important, it also reflects major changes in how electricity is used and where it is generated. The key technology to unlocking the economic value presented by this new market is analytics, coupled with an understanding of both building technology and electricity markets.

As a founding member and chairman emeritus of the U.S. Department of Energy GridWise Architecture Council, the author had the opportunity to play a role in shaping the evolution of this new electricity market and the technology that has come to make up the smart grid. As an energy manager for more than three decades, participating in development of BAS, open systems, energy management and data standards, the author also saw the evolution in those disciplines and was able to apply this knowledge to the smart grid. This combination of experiences has been ideal for positioning the author to understand this rich landscape of opportunity and share it with the reader. However the landscape is even more complex because the energy, automation and smart grid evolutions occurred during a time of huge growth in two other areas—electricity and the internet. With electricity, onsite electric generation from renewables and conventional sources has grown dramatically. The onsite generation trend was driven by a series of factors, like climate change and energy resiliency after a series of "super storms." Starting with the massive power outage on the eastern seaboard of the U.S. in 2003 and continuing with climate change and resiliency issues, these issues with electricity resulted in wide ranging new government policy that shaped the development of this new electricity marketplace. Without effective access to and use of information, or analytics, it is not possible for managers to leverage the full economic value of technology-enabled building/energy management and onsite generation that are the hallmarks of this new electricity marketplace. Understanding how to capitalize on this new marketplace requires fluency with the technology, which is the purpose of this book. More importantly, this book is about the power of the information that resides in buildings, all of the new electricity technology, and how that information can be manipulated in an automated fashion to enhance operations and optimize the economics associated with running an enterprise.

This chapter introduces the overarching topic of buildings and energy/electricity management along with the power of analytics. The evolution of the technologies, systems, and business rules that govern the marketplace presents both opportunities and challenges for managers. These managers realize that it is critical to integrate systems for seamless facility operation, but also to provide access to lucrative utility markets that offer significant cost savings for smart buyers. Part one of this book will begin to outline the new *electric reality* that will be referred to as $4^{plex}$ *energy systems*. Energy, building and campus managers are among those professionals who face daunting challenges as they work to operate increasingly complex facility energy systems.

The author coined the term *"$4^{plex}$ Energy Systems"* to describe the new electricity environment, and analytics is a key topic because effective systems operation requires sophisticated analysis. Managing these systems orchestrates power and energy use from four sources:

1. Power from local utility,
2. Power purchased from electric retailer,
3. Onsite generation from solar and other systems,
4. Efficiency *PLUS* demand response and market participation including bidding in synchronized, day-ahead, and real time markets.

Chapter 2 will provide a more detailed discussion of $4^{plex}$ energy systems.

With regard to the new energy marketplace, the reader might ask what are analytics and why should we care. These topics will be explored in detail, but quite simply, analytics are software systems that perform analyses on operational information, in the form of electronic data from many sources. The analyses compare actual operating data to "rules," which indicate ideal operating conditions. The result of these analyses is that analytics change data into knowledge. This knowledge allows a system manager to evaluate performance and to take action to improve performance, creating economic value. The energy and analytics payday is one reason this topic is so important. This payday comes from efficiency cost savings that are achieved through optimization of energy use, but it can also come from the cost benefits available from participation in the deregulated electricity market. In this market, legislative and regulatory policy has been enacted that requires utilities to pay energy users for changing electricity usage patterns through demand response (DR) and to buy electric capacity that is offered to the grid. Building and Certified Energy Managers (CEMs) can utilize sophisticated building automation systems (BAS) to participate in these programs. They can also use power from onsite generation to operate more cost effectively, and to sell power back to the grid.

To truly leverage the power of analytics, readers must develop a good working knowledge of a significant number of technology topics that are covered in this book. With the advent of smart grid and the growing predominance of high-speed data communications in most homes, much of this technology is being deployed residentially as well. That is why Part II of this book will provide a thorough treatise on these technologies. Among these technologies are building automation systems, data communications and networking, and "Middleware." Pervasive standards for data communications, such as BACnet, are the price of entry for buildings to be active participants in new energy markets. Middleware involves a new class of technology products, which allows owners to "integrate" legacy BAS systems with new standard systems, but also opens access to a host of other system integrations.

The reader may already be aware that building systems are made of a set of complex technologies which in many cases operate independently. The movement toward integrated, or smart connected, systems via middleware and web access presents that opportunity for a wealth of building information. Yes, this information is disparate, coming from many independent systems, and it comprises huge volumes of data that can be daunting as well. The first task is to distill the volumes of data into knowledge. For example, 15-minute-interval data on electric demand from a utility revenue meter just looks like an array of numbers. An energy manager's skilled eye can make sense of those data, but these professionals don't have time if it is necessary to review tens or hundreds of meters, each with 35,000 interval data points per year (4-15 minute reading per hour x 24 hours x 365 days). Analytics software, on the other hand, can analyze those data to determine when peak consumption occurs and whether those times are appropriate electronically. With this information it is possible for the energy manager to focus on anomalies that may be pointed out by the analytics. Energy and analytics tools have been developed to sit at a higher level of the architecture, above the building systems, to access data from those systems, and to perform analyses. Again the volume of data is daunting, so these systems can identify problems and see opportunities that the manager cannot identify in any other way.

In the past, a major obstacle to deploying this type of building analysis has been the cost associated with implementing instrumentation at every piece of equipment, deploying data communication networks to transport those data, and finally deploying a highly sophisticating processing engine to conduct analysis. All of those factors translated to very high cost. Energy and analytics as discussed in this book sit at a nexus of technology development that has made it possible to overcome all of those past challenges and to unlock the value that the technology can offer. Cloud computing, the internet, data communication standards, and web services are among the 21st century technologies that have removed the obstacles to implementation of analytics. Through this book analytic architecture will be discussed in detail to clarify systems technology options. This is particularly important as managers consider integration using middleware, web services and software as a service (SaaS)—alternatives that are common in the world of analytics. All of these options create specific requirements for analytics technology, making it essential for this book to discuss them along with the evolution of standards for networking and the internet.

Case studies highlighting successful analytics applications will be covered in Part III of this book. Common applications for this technology were discussed earlier in this chapter, and this book focuses on energy and analytics. Professionals are demanding tools and technology to drive positive energy results in this complex business and technology environment for buildings and campuses. In the face of that demand, the market has responded with a barrage of evolving hardware, software, and web services, making decisions about which analytics tools are best challenging. This book will address that topic under the chapter on analytics tools, but will also provide case studies showing successful implementation of the technology. Managers want to achieve optimal energy efficiency and also get the maximum useful life from current equipment and technology investments. There is no better alternative to achieve such goals in today's market than energy and analytics.

# Chapter 2

# 4ᴾᴸᴱˣ Energy

The energy commodity business (sale of units of any type of energy) has seen a number of technology waves and engineering evolutions over the decades. Whether the reader is a building operator, Certified Energy Manager (CEM), engineer or other professional dealing with a single building/campus, overseeing multiple buildings, or a consultant keeping pace with technology that is barraging the industry is a *resource* challenge.

As discussed in Chapter 1, energy and analytics encompasses a wide range of technology; however, the focus for technologies to be discussed here is on energy and buildings. Energy and analytics offer a wide range of application functionality, and these technologies typically include the analytics algorithms, as well as dashboards to visualize reports. This overarching topic of energy and analytics is a monster trend impacting energy, buildings and large-scale energy management across one or many facilities. Considering that approximately ±70% of the energy consumption in buildings is electricity begs the question, how can energy and analytics technology support managing and reducing that consumption? The answer to that question may become clear if that discussion emphasizes the *"EcoImprint" (EI)* of buildings and campuses. EI is more than a carbon footprint, which implies local impact only. EI takes into account the far reaching impact of electricity. The imprint for electricity, as an energy resource is unique, because it leaves very little imprint where it is used, rather it leaves an imprint tens or hundreds of miles away where it is generated, transmitted and distributed. The imprint also extends to where the source energy (i.e., coal) for the generation is extracted as well. A major driver for President Obama's Clean Energy Plan, which was announced by the Environmental Protection Agency (EPA) in June of 2014, was the realization that coal-fired power plants have a significant carbon footprint. This is therefore about a far reaching resource impact, but eco in this case has two meanings. First it refers to the <u>eco</u>logical (environmental and climate) impact that the full electricity system has on land, air and water, but it also refers to the economic impact of electricity. Electricity is a commodity for which the cost can vary from pennies a unit, to thousands of dollars per unit, depending on customer use at different times of the day, month or year. Highlighting the ecological and economic impact of electricity is critical. However it is equally important to also understand that this energy source is essential to quality of life, business vitality and the economy as a whole in the 21st century. Leveraging technology to maximize the output for every unit of electricity consumed (resource efficiency) and minimizing the ecoimprint of every unit consumed (economic efficiency) must be the cornerstone of planning the future. The technology tools provided by energy and analytics, and visualized by dashboards, are ideally suited to fulfilling those tasks.

Technologies in the form of numerous energy and analytics tools, which may include software and services or solutions, are being presented to CEMs, building managers and consultants on what seems like a daily basis. These technologies could be called "disruptive" because they are changing the way energy management is carried out. It is also worthwhile to note that analytics is not new technology looking for a market; on the contrary, there is growing demand for solutions. The demand is from CEMs and a range of building and campus managers facing real energy challenges, as they work to operate increasingly more complex facility energy systems. Beyond system complexity, these managers are also striving to control energy cost in an ever more complex marketplace. Energy management given these challenges has become more challenging, but there is another new dilemma, energy resiliency or "keeping the lights on." In the face of "super storms," attributed by many as a real indication of climatic impacts from climate change, essential facilities of all types are grappling with the effects of long-term power outages and looking for ways to protect themselves.

The wave of emerging technology for energy and analytics is still very new to energy and building professionals. The intent of this chapter, and this book, is to provide an introduction to next generation energy technology and to describe how it is being used. Some examples of past building system technology evolutions that required energy engineers to embrace new technology were: the move from mini-computer based building automation systems (BAS) to microprocessors for BAS with distributed direct digital control (DDC), sometimes called energy management systems (EMS); BAS legacy systems and proprietary protocols migrating to data communications standards like BACnet™; and then middleware (i.e. Tridium), Convergence, Web Services, the cloud and now analytics and dashboards. These evolutions occurred over four decades, and the same managers were also dealing with a host of other technology evolutions at the same time including distributed generation technology ranging from solar or combined heat and power (CHP) to microgrids, advanced energy metering, deregulation of natural gas and electricity leading to energy procurement, demand response and much more.

The notion of $4^{PLEX}$ $Energy$™ takes into account the full range of complexities outlined above, with a focus on technology that can assist managers in addressing energy challenges in today's market and beyond. In fact, this nexus of energy and technology may have motivated the reader to pick up this book. The reality is that energy markets and technology are in a continual state of flux, so this chapter cannot anticipate every evolution that might occur over the years. Yet within the context of the $4^{PLEX}$ energy segments covered here, it is the author's belief that this material will remain timely well into the future. This discussion, by defining the concept of $4^{PLEX}$ $Energy$™, and describing how each aspect of it relates to energy and analytics technologies, as well as visualization tools or dashboards. Dashboard will be discussed in some detail in a later chapter. This book has several chapters devoted to the use of analytics for energy management, building operations and managing electricity, but prior to discussing $4^{PLEX}$ $Energy$™, it may be helpful to begin with a brief treatment of energy and analytics.

## ENERGY AND ANALYTICS

The technology discussed here might be called *"Big Data for Buildings,"* or *"Big Data for Energy,"* but this effort has elected to focus on *"Energy and Analytics."* This is because the author believes that emergence of this technology could not have been more timely, in light of the significant opportunity for improvement that it offers for energy, building and campus management. For CEMs, building owners, and managers there is a critical need to optimize performance, to control cost, and to find new ways to leverage greater value from buildings as assets. For the buildings community, energy and analytics is exciting because it is a technology-rich discipline, which requires professionals who understand every aspect of building operations, as well as the complex technology deployed. Building technologies and activities that can find value in analytics include; commissioning, metering, building automation systems (BAS), dashboards, middleware, web services tools for buildings, energy services. Of equal value is that each of those technologies and activities is also a source of data for energy and analytics. With content knowledge in all of these areas, the reader will be better positioned for analytics success than anyone else in the building space. This is combined with an understanding of energy management programs like United States Green Building Council (USGBC) LEED™/PEER™ and EPA Energy Star™, as well as utility efficiency rebates, continuous commissioning, meters, etc. Figure 2-1 is a bit busy, but it should provide insight into how these technologies, combined with the 21st century electricity market offer exciting opportunities for integrators and building owners.

Readers might associate analytics with "continuous commissioning or "fault, diagnostics and detection" for HVAC, but as outlined in Chapter 1, that is just one of the applications for this technology. Analytics, again focusing on energy in this book, are generally combined with visualization tools, or dashboards, to provide an essential management tool. This tool can unleash value by optimizing building operations and energy management, particularly electricity, use and cost. It has been noted here that electricity is the highest quality form of energy, and it is essential for a 21st century quality of life, including education, business and entertainment technology. Yet electric energy cost represents the largest *controllable* operation expense in commercial buildings, and the U.S. Department of Energy (DOE) has predicted that these buildings will make up 80 percent of projected growth in energy consumption between now and 2040—a good reason to leverage energy and analytics technology for energy management.

ENERGY MANAGEMENT APPLICATIONS FOR ANALYTICS

Analytics present exciting technologies to enable energy managers to participate in an electricity marketplace, which offers huge savings opportunity through ecoimprint reduction. The author calls this new market 4PLEX Energy, and it is a central "application" for analytics. Analytics is the brain that enables both participation in and value genera-

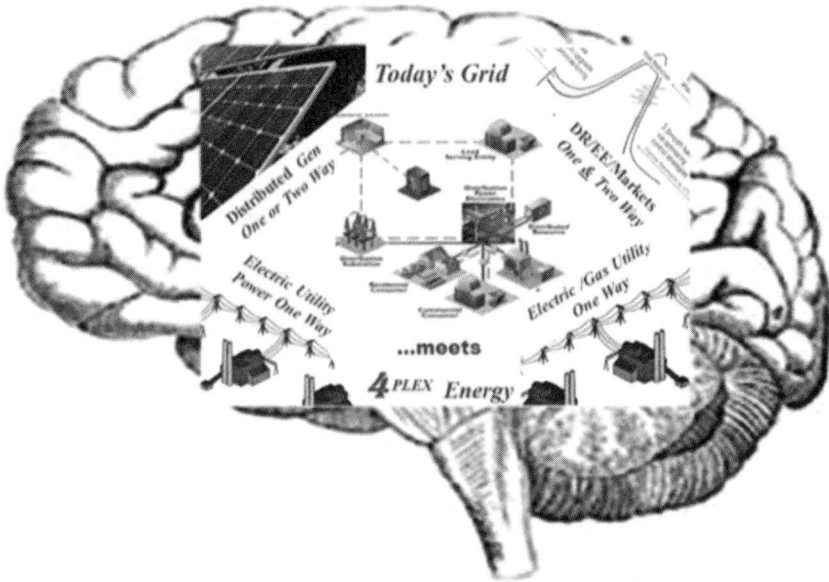

**Figure 2-1. Analytics are the Brains of 21ˢᵗ Century Building/Energy Management**

tion from this market. A simple explanation of how analytics tools work is:

- Analytics contain "rules," which define an ideal performance metric for equipment, etc.

- Analytics are deployed with data communication technology that enables them to access current operating data from equipment, etc. and

- Analytics execute algorithm-based analyses to compare current operating data against rules, thus determining current performance.

Based upon the analytics applications discussed in Chapter 1, this process can determine performance for a wide variety of processes. At times the term key performance indicator (KPI) will be used to refer to the performance metric of interest. In this book, performance metrics will primarily focus on energy management, but as discussed in Chapter 1, the basic analytics process is somewhat universal. Electricity is a complex and data rich environment that the book will cover in detail, because understanding these concepts is necessary to achieve energy and cost savings. As the reader considers the analytics process outlined here, and the real time, fast-paced, business of buying, using and managing electricity, these savings benefits should become more evident.

With an introduction to electricity and analytics, this discussion will now focus on the concept of $4^{PLEX}$ Energy. As the name implies, this concept highlights four types of energy. The first $4^{PLEX}$ Energy type has been around for more than 100 years—electricity purchased from the local utility. It has often been said that the utility system was designed for *energy to go one way* and *money to go the other,* when the bill is paid. The electric meter is generally seen as the demarcation point between "supply side" (generation, transmission and distribution of electricity) and "demand side" (the customer's use of electricity). It may appear at first look that there is little new in this plex-segment of electricity, but look closer. The supply side of this segment is dramatically different in deregulated states and, with implementation of the President's Clean Energy Plan, it could become even more diverse. Many of these changes will be discussed under the fourth plex segment below, but it is clear that the electricity business is diverse, complex and in a state of change. One of the most complicated factors about this business is that it is regulated independently in each state. The Federal Energy Regulatory Commission (FERC) regulates the wholesale electricity business, and in recent years has had a major impact on retail markets in deregulated states, but the retail electric business is still regulated, for the most part, by each state public utility commission. This adds complexity for building owners who operate in multiple states because they must analyze the business rules independently in each state before applying energy and analytics technology.

The second $4^{PLEX}$ Energy type exists because electricity is deregulated in approximately half of the U.S. states, and natural gas is deregulated nationally, so either fuel may be purchased from a retail energy provider (REP). The basic transaction looks the same to a customer, and again energy goes one way and money the other. The premise is that "unit cost" for a kilowatt hour of electricity is lower when purchased from a retailer, but it still must be delivered over the local utility's distribution system, often call "wires." Therefore the customer pays both an energy unit cost and a transportation or wires charge for use of the utility system. Buying electricity this way will make sense for a customer if the net cost of energy and distribution from the REP creates a savings compared to buying from the local utility. It is also worth noting that the distribution fee is an added cost, and generally the customer will still get a utility bill for that fee. These fees are established as part of "rate cases" held before the state public utility commission (PUC). When rates are approved by the PUC, a utility has a fee structure that it may use to bill customers for energy and for fees, such as distribution charges. State PUCs establish distribution charges that the utility may assess on electricity units purchased from retailers (where deregulation exists) and for distribution of natural gas energy. Analytics offer an excellent tool to analyze energy consumption profiles and create metrics that can help managers negotiate the best pricing on energy procurement.

The metrics available from energy and analytics tools also enable managers to use the energy consumption profiles to identify efficiency opportunities, and to better understand how to plan for $4^{PLEX}$ *Energy* types 3 and 4. The reader should note that if this content is not completely clear, the purpose of this book is to offer further discussion that should provide clarity on all of these topics. Energy profiling can be approached in a variety of ways, from simple spreadsheets that track consumption, demand and also summarize costs for each as well as service charges, etc. Utilities can be very helpful in this area by providing online tools to access billing histories, and for a nominal fee, they will install an interface to allow the building owner to connect the meter to a building automation system (BAS). These data are very helpful in tracking managing and rectifying bills, and to understanding electricity consumption profiles. The most sophisticated systems will enhance this capability by capturing "interval data," which captures and tracks information at 15- and 30-minute intervals. This is particularly valuable to understand peak demand and correlate equipment operation with consumption profiles to determine whether cost savings can be achieved. These data can also be very helpful in negotiating final cost proposals from electricity retailers.

The third $4^{PLEX}$ *Energy* type is distributed generation (DG). Generating electricity onsite can be a highly effective way to reduce cost, improve sustainability and electricity "resiliency" or reliability and much more. It is not the intent of this discussion to review design criteria for sizing the right amount of onsite generation. However it is important to utilize interval data and analytics to ensure that DG has been deployed effectively, and that the overall operation of the facility is driving the greatest benefit. This type of $4^{PLEX}$ *Energy* has seen significant growth over the past two decades, because there has been extensive implementation of DG from solar photovoltaics (PV). PV has been deployed at utility scale, but has also become very popular for installation onsite by customers. Solar caught the national consciousness and has been installed at all sorts of commercial and residential facilities. The popularity has been stimulated by federal and state subsidies and a variety of business models like power purchase agreements. Dramatic decline in the cost of solar technology has further stimulated installation of renewable electric generation. That is all good news for the energy consumer, but utilities feel very differently. Their big problem is that consumers who generate power onsite are taking money out of their pockets, and they also complain that the output in electricity from solar systems is variable. The utilities say that they still need to build power plants to ensure that power is available when the output from the solar system varies downward because a cloud blocks the sun. This is a complex topic that is beyond the scope of this section, and it remains to be seen if these issues will result in market changes that reduce adoption of solar technology. Readers who are interested should pay particular attention to what is happening in the local market. Yet renewables are not the only form of DG. There are still many buildings, often campuses, with combined heat and power systems (CHP), traditionally called cogeneration. With deregulation, and the ever increasing number of "super storms," cogeneration has seen resurgence. In many cases, these systems are now being operated through "microgrids" that are capable of interaction and trading of power with the local utility. After Super Storm Sandy, for instance, large parts of the eastern seaboard were without power for days. Notable exceptions included places like Princeton University, which was able to insulate its facilities from the outage that occurred after Sandy because it had distributed generation and operated a campus microgrid.

*Energy resiliency* has already been mentioned as a term that is used to describe actions that facility operators are taking to protect themselves from long power outages after major storms. Of course, utilities are taking action to make the grid more reliable and resilient, but with growing demand and the potential that supply-side assets like coal-fired and nuclear power plants may be taken offline, this issue is creating more concern among electric consumers. Actions by consumers typically involve distributed generation, though there is growing viability in new forms of electricity storage. While motivation for most renewable energy installations focuses on sustainability, customers who build other types of DG are usually more interested in creating resiliency by building a microgrid. These DG systems are more likely to use conventional energy sources like natural gas. The author expects to see more and more hybrid systems in coming years, which combine renewable energy and conventionally fueled electric generation technology. In any case it is typical to size these systems to provide part, not all, of a building's energy needs. Analytics are essential to the design of these systems for several reasons. First, the analytics tool maintains energy profile data, based upon utility bills, that are ideal for determining the right balance between utility-supplied power and onsite generation. Second, it is common for analytics tools to use "middleware technology" to access "interval data" from meters, BAS and other sources, which provide even more meaningful data on energy profiles that include energy (kilowatt hour) and demand (kilowatt). Even more valuable in this process is that the analytics tool can use the same middleware to develop reports that inform the customer what loads (equipment using electricity) are operating during peaks throughout an entire year. The resulting value this energy and analytics process offers to the manager is unparalleled to achieve optimal system sizing. Even more exciting is that the manager can use this information to develop truly 21$^{st}$ century energy management programs that leverage building automation system (BAS) technology, energy procurement from the utility and the REP, DG and programs that will be described under the fourth type of 4$^{Plex}$ Energy. Developing these robust energy profiles is extremely important during a major electricity outage, because it enables the managers to use the BAS to shut down non-essential equipment, so that they can operate on less power. During all other times though, managers can operate this equipment to save money through efficiency and to reduce utility cost. The building and energy manager should recognize that this may sound simple, but it is a completely new discipline for many operators. Traditionally, building operators have focused on equipment operations, and maintaining a building to ensure that occupants have ideal conditions to carry out their activities. For example a hospital surgical suite has a different profile for temperature, humidity, etc. than an elementary school classroom. Fulfilling these tasks was already daunting for building operators. Now asking a building operator/manager to take on the added responsibility of understanding and managing the use of all building equipment to meet energy/electricity consumption levels is a new and complex task. That is why analytics tools bring tremendous value to the operator and those who ultimately benefit from the savings that result from these new complex tasks.

The fourth $4^{PLEX}$ *Energy* type is really about monetizing the value of energy, particularly electricity units, which the building does not consume. With effective analytics tools, at certain times managers can actually sell back to the grid power that they do not use. This means that electricity and money can both go two ways. This process is governed by a variety of programs, and when the building/manager participates, they can get a revenue stream from selling power (that they do not use or that they generate onsite and send to the grid) back to the utility.

The term "monetizing energy" may seem foreign to the reader, but it is worth learning about. Again, these programs are for electricity, and the benefits may apply in both regulated and deregulated electric markets (states). The basis for this idea is that electricity is a unique commodity because it is generated and consumed almost instantaneously. As the reader knows, electricity cannot be stored in a warehouse for use later. For that reason, the value fluctuates dramatically over time, based upon how much electric demand there is from customers at any specific time. Many may say, "Wait, my energy bill each month shows a number of units or kilowatt hours (kWh) consumed, and every unit is charged the same unit cost." That is true but the nature of the monopoly, regulation business model makes it possible for utilities to make up the cost volatility in various ways. One way is for utilities to add demand charges for large users, which assess an added penalty to those who consume more during high use periods. Other charges may be assessed as well. The real point is that the information on the utility bill can be confusing and only tells part of the story. Again, electricity is a regulated monopoly business and rate structures (prices) are established to insulate the consumer from short-term fluctuations in the market. At the same time these rate structures also allow the utility to recover its cost of operations, as well as the cost it incurs from building new electricity infrastructure (generation, transmission, distribution, etc.). The intent of this text is not to provide a treatise on utility rate analysis, but to point out a simple fact: The cost of electricity is changing continuously. At the state level, retail pricing is regulated by public utility commissions, but utilities invest tremendous legal and staff resources to ensure that all of their costs are recovered, plus a profit margin, through that single "rate" or price per kWh. The more important point here is that the US electric grid is under more pressure almost daily to keep the lights on in the face of ever growing demand for this precious commodity. Many of the appliances, devices, etc. that we associate with business excellence, and quality of life at home, are completely reliant on electricity. The net result is that utilities have had to develop a whole series of options (programs) that enable them to balance supply-side resources with demand from all of its customers. In the old days, utilities forecasted future demand at regular intervals and requested rate increases from the PUC to cover the cost of building more and more new infrastructure. In the 21st century, a host of factors that have been discussed obstruct this old approach to doing business. Utilities will still build new infrastructure, but under pressure from customers and regulators they are also developing programs to use demand from customers. Again customers who participate in these programs are rewarded monetarily for changing consumption profiles. Changing those profiles relieves demand from the grid and makes sense as a strategy to reduce the ecoimprint of the utility and to hedge against the rate at which new infrastructure must be built. The program rewards customers for changing behavior/electricity consumption, and in many cases these payments provide the incentive for customers to invest in efficiency and renewable energy projects. This is what has been called monetizing electricity. Critical to this discussion is that commercial buildings cannot monetize value from such changes in behavior on a repeatable basis without building intelligence systems and analytics.

This fourth type of *4Plex Energy* includes a rich ecosystem of programs and strategies that will pay customers to change behavior and consumption patterns in return for payment. Again analytics are the price of entry to maximize the benefits of such programs, however, and a robust system of building intelligence technology is ideal as well. A great example of these utility programs, which excel with BAS and analytics technology, is demand response (DR). It is also exciting that the same technology, which enables DR, can also be used with capacity, reliability, bidding and other programs in deregulated electricity markets that allow customers to sell power back to the grid. Again this may be power that the customer agrees not to use for a period of time, or DG power that is put back onto the grid through an "interconnect" agreement. All of these programs are based upon the time value of electricity, and customers must contract with the utility through an enrollment process of some type. Concurrent with the enrollment, customers will also install technology to enable these strategies, which will be discussed later in this book. These strategies allow consumers to be paid for changing electricity consumption patterns, and for implementing methods to sell power back to the grid. Blending these strategies with microgrid DG creates even greater opportunity for monetization.

In a $4^{PLEX}$ *Energy* market, energy managers and all professionals will benefit from both understanding and leveraging the technology discussed here. Electricity markets and programs offer real economic value, but applying analytics is the ideal way to unlock those benefits for building owners. There are also requirements in many markets to implement automated demand response (ADR) technology to execute the sequences. Taking this a step further to also couple BAS technology with the power of analytics software (i.e. Skyspark™ or Analytika™) can create even more value.

Readers may want to avail themselves of the many opportunities that exist to gain knowledge in this area. Local utilities usually offer training for many of these programs. Trade associations like the Association of Energy Engineers, *www.aeecenter.org*, are great sources of training and information provided through conferences and seminars. The reader should also pay attention to opportunities for seminars and articles in industry periodicals as well. Project-Haystack is a relatively new organization that focuses on helping drive standard approaches for sharing device data among applications, such as analytics and visualization tools. The open source community hosts an event called Haystack-Connect on a bi-annual basis—*www.haystackconnect.org*. These are just a few of the great places to learn more about this exciting area.

$4^{PLEX}$ *Energy* is an indicator of how dramatically energy management is changing. Taking advantage of the technology, strategies, programs, etc. outlined in this chapter can drive huge value for managers. Further, the author believes that a voracious appetite for information about energy and analytics, as well as electricity markets, will pay off with exciting opportunities for buildings and businesses.

# Chapter 3

# Dashboards and
# Visualization Tools

In 2001 this author introduced the term "dashboard" to the energy and buildings community. It was not a flash of insight, but rather the awareness that a trend, and a term, which was already established in the information technology (IT) world was appropriate for this space. This chapter will focus on the topic of dashboards as a "web-based interface" for building systems including building automation systems (BAS) and myriad technologies from analytics to metering, onsite generation via renewables and cogeneration, special systems such as security and life safety, digital signage and a host of other computer-based equipment.

The focus here is on visualization and dashboards, but it may be helpful to revisit energy and building technology network standards, and the underlying basis for enabling internet-based interface. An entire chapter is devoted to this subject later in this book to provide detailed content for the reader. In the late 1990s and early 2000s, BACnet and LON were vying to become the data communication and networking standards for building automation systems (BAS). By that time it was already clear that building and energy managers wanted technology to unify the interface to these systems. In fact, they were demanding a single-user interface to all BAS, and this group vocalized that demand starting in the mid-eighties. It was not uncommon then, and even today, for building operators to have a room, or a countertop, with multiple computers, each running different BAS and building systems software interfaces. What this author, and others, started to write about and to demand in articles, papers, etc. beginning in the early 1980s, ultimately became known as "open systems." A standard and/or unified BAS interface was considered one of the primary goals of open systems. As early as the mid to late 1970s there were companies offering aftermarket systems for unified BAS interface, but the technology was immature. A secondary demand from owners began to take shape in the late 1980s and early 1990s, and this was for an open systems approach to BAS expansions and would allow them to "integrate" BAS from multiple manufacturers. As a matter of practicality, the standards committees ultimately made a decision that this latter need for "integration" was the highest priority for open systems. As a result, the open systems standards development process became focused on the data communications networking for BAS systems, and the services that have to be addressed to operate a distributed direct digital control system. The reasons for this were many, but at its heart was the belief that the cost for systems is in hardware and software that provide the automation, and the best way to control system cost was to make it easier to integrate existing systems with new ones. The key complaint from customers and operators was that they were "locked in" to a system because the automation technology, including the equipment and the data network, was proprietary. The interface was also proprietary, but it represented less cost. Fast forward three decades and that decision looks like pure genius. Why? Because of the internet and ubiquitous web interfaces.

Merging these web interfaces, or dashboards, with analytics, even embedding the dashboard within the analytics, is the next monster trend that will impact energy, buildings and BAS. At one time, the computer interface was a differentiator for BAS manufacturers, but now many of these companies are integrating third-party dashboards with their systems. A variety of products utilizing variations on these technologies are being presented to energy engineers, building managers and consultants on a daily basis. These technologies might even be considered "disruptive" because they are changing the way building and energy management is carried out. There is actually great demand for solutions to a host of day-to-day energy management challenges, which these tools are targeted to meet. However, this wave of technology is new to many energy engineers. The intent here is to provide an introduction to this next generation dashboard technology, and describe how it is being used. This technology is evolving rapidly and no book can keep pace with all of those developments as a buyer's guide, but it is possible to introduce fundamentals of the technology that will assist buyers in asking the right questions. Dashboards and analytics will be somewhat disruptive in the way that they impact energy and building management, making it critical to include these issues in any long-term operations plan.

## DASHBOARDS

Since being introduced to the energy and buildings community, the term "dashboard" has reached "buzzword" status. Yet a dashboard is nothing more than a software visualization tool to provide system information about a facility. The IT industry was already applying dashboards to all kinds of applications decades ago. Consider how Amazon. com tracks buying preferences and offers up the kinds of products a consumer is likely to buy. With the proliferation of BACnet™ systems, and the advent of web services, it became possible to explore new visualization technology in this space. As mentioned above, prior to this dashboard explosion, building and energy managers often had multiple computers, each running different BAS software interfaces. This is still the case for many users today because dashboards may be available today, but the technology is new and intimidating to some managers. As a result, one purpose of this book is to assist those professionals with understanding and making good purchasing decisions regarding this technology.

The genesis for dashboard technology in energy and buildings may have been demand for a unified BAS interface, but in fruition these systems are unifying more functions than just BAS. These functions or systems include those mentioned in Chapters 1 and 2, which in the past required independent computer infrastructure, data communication networks and ancillary gear for visualizations. Among these systems may be: computerized maintenance management systems (CMMS) or continuous commissioning technology, metering, and many more. In this chapter, a sampling of implementations will be discussed, but what is even more exciting is that integrating these previously disparate systems also opens up the true power of big data for energy and analytics. For example, integrating BAS and metering systems for interval data, along with a utility interface dashboard, makes it possible to achieve efficiency and electric demand savings on the customer side of the meter. This concept was covered under $4^{PLEX}$ *Energy* and presented exciting opportunities for energy management.

In the 21st century, almost all BAS manufacturers offer some form of web-based interface, or dashboard, for their systems, but a robust aftermarket of dashboards exists as well. These dashboards have been widely accepted, and a robust web services market has developed from companies that are offering customized dashboards to the BAS manufacturers. This means that BAS manufacturers have come full circle, from using the software interface as a differentiator to accepting that they are now ancillary products sold to the automation industry by many suppliers. Quite simply, these dashboard tools are truly "interfaces," meaning that they may be used to monitor the BAS, but not to configure or program the BAS components or networks. This is a fact that energy/building managers have accepted, along with the understanding that a separate software tool is needed, and must be purchased from the BAS manufacturer, to complete configuration of systems and to program sequences of operations on any control device. There are also a growing number of "workbench" tools being introduced to the market, which provide non-manufacturer developed configuration and programming tools. The viability of such tools, however, is limited by the accessibility of operating systems, data communication networks and programming tools to external developers. Of course technical viability is not the only issue; the greater concern is whether or not there is a market for programming tools. As with all products, a market opportunity would only exist if the tools offer unique features that deliver real value to customers. It remains to be seen whether such features will be offered, but what is clear today is that dashboard interfaces are generating tremendous interest and creation of a whole new eco-system of companies in the industry.

The fact that dashboards cannot configure and program the BAS, or other building systems, is not necessarily a limitation, because the true value is in using these tools for energy and building management. All of the tasks outlined in the last three chapters for equipment operation and energy management require dashboards. The dashboard is also critical as an interface to the analytics tool allowing managers to develop and track consumption profiles, conduct FD&D, etc. Given this distinction between programming tools for initial BAS setup and dashboards for day-to-day operations and energy management, managers can decide whether to have in-house programming expertise on both types of systems or to outsource one of these functions. After initial startup, there is limited need for the programming tool, except when it is necessary to replace a controller or tweak program sequences as necessary.

The term "dashboard" now encompasses much more than BAS web interfaces, but these interfaces are critical to successful long-term energy management. Coupling the dashboard with analytics to evaluate specific KPIs is one way that managers can trump the single greatest cause of BAS failure. That happens when minor changes and "overrides" to equipment program sequences occur little by little over time, until the systems are completely ineffective. Normally no one at the building even notices that systems are being circumvented, but the *analytics tool* will notice. This chapter will provide a situation analysis on the dashboard technology, as well as on the dashboard applications.

## DASHBOARD TECHNOLOGY SITUATION ANALYSIS

Dashboard technologies will be classified in two broad categories for this discussion: 1) operational tools for energy/building managers, and 2) kiosk tools for building occupants and visitors. In fact, both are primarily software tools that leverage data from a host of systems onsite and online. The earliest incarnations of operator tools did not see widespread implementation because they required a hardware system like a computer, along with an extensive network of ancillary components, which had to be hardwired back to a BAS computer. The first cost was prohibitive and few facilities, other than universities and industrial applications, had the staff sophistication to use them, as well as the energy and operating costs to justify the investment. Today, operational tools are lower in cost to implement and much more user friendly. Implementation costs have been reduced due to web-based data communications standards, beginning with BACnet™ but including many others, and these systems have incorporated web services compatibility.

Communication standards, and web services, combined with expert system integrators who can create (program) drivers to access data from "legacy" or proprietary BAS, and other equipment, have launched an independent world of "middleware" products. Capable integrators can buy these products and program or purchase "after-market" drivers that make it possible to communicate with, and extend the useful life of, legacy BAS equipment. This is because the equipment can be maintained in operable condition at the building, or even application, level so that the manager can develop "migration paths" to allow phased upgrade of legacy to standard systems over time. The take-away for dashboards is that all of this technology can extract the data needed to provide an interface to almost any building system. The Ethernet LAN, VPN or other network that already exists in most buildings, and on most campuses, is able to move those data, and they can be housed on a local computer, or a server accessible via wide-area network or "the cloud." This marriage of technology and need has opened up a new world of functionality and dramatically reduced the cost of the systems. That means that almost any managers' requirements for access to energy and building data can be met by today's dashboards.

Given this complex building level, data intranet, a dashboard can provide BAS interface to get status information about a building or campus. Dashboards combined with analytics can then be expanded to provide the features of enterprise energy management or continuous commissioning. These latter features are disciplines within themselves, and will not be discussed in detail here. Twenty-first century energy management, discussed in Chapter 2 as $4^{PLEX}$ $Energy^{TM}$ highlights some of the ways these technologies can bring value. Dashboards discussed thus far are software driven tools, used by operators, typically for two reasons. First to optimize building performance and operating cost, and second to ensure that the building is maintained in a way that serves the mission. Building occupants pay no attention to the complex equipment that maintains the environment, but if a surgical suite must be maintained at a specific temperature, or an office building is uncomfortable, major issues develop very quickly. So the building operator's job is to manage to a set of requirements that defines the optimal building environment, and work continuously to meet those requirements. Cost is an ongoing factor in any operation, so the true value of a dashboard is to give an operator one tool to use for monitoring the building and the output of the analytics tool. In this case monitoring is a real-time activity that operators carry out attempting to address issues that arise, and to ensure that the building is operated within specific tolerances. Those tolerances may be comfort or process, but the management process must also control cost and energy consumption. With the growing emphasis on sustainability there are mandates and policies that require buildings to be operated to a number of indices. To be successful these activities must be integrated management activities combining the power of the BAS, analytics and dashboard with the content knowledge of the manager. Operators are only capable of conducting so many tasks at one time, and particularly when conditions and variables are in a state of continual change. It is not possible to do this work with a dashboard alone, rather analytics become essential. Determining the optimum role for dashboard interface becomes very important to the energy/building manager, and these tasks will become clear when we discuss analytics.

The second type of dashboard will be called a "kiosk." In recent years it has become very popular to summarize high level building data on a screen that is available for any building visitor to see, or available to managers and others within an organization. Providing this sort of user friendly interface is a great way to share sustainability accomplishments with building and campus users. Often a kiosk or touch screen is provided in a building entrance or other central area to summarize high-level information about sustainability and efficiency goals, as well as current or real-time performance against those goals. This same level of data may also be duplicated on a "cloud" based website, or as a link from the organization's website to make the information available to a very large audience. The data provided on a physical screen or kiosk, typically are produced by a server, or operator dashboard, that resides elsewhere on a network. The data are then "pushed" to a computer/computer screen for display to the visitor.

Kiosk dashboards are usually graphically driven, much like a PowerPoint presentation, and may incorporate other information for visitors like an office directory, electricity output from an on-site solar array or fire evacuation plans. These are sometimes also referred to as digital signage and can include the day's events or even advertising. Ultimately the primary value of dashboards is awareness about current status and performance of the building or campus, which is easy to visualize and understand. There is very little analysis provided by the tools other than perhaps some averaging of information points, etc.

This was a high level discussion of dashboard technology and the reader may want more technical detail. To address that need, this book has been separated into two parts. Here in Part I the trends, business reality and overarching technology topics related to energy and analytics are covered. To better understand the underlying technology, Part II of the book will dive more deeply into the technologies themselves. A curious fact however, is that the underlying technology requires study of multiple related topics. Those topics include: BAS and other systems technologies, data communications and networking technology, middleware and more. This book cannot provide detailed training on computers, or information technology, so the author will assume that readers come to the topic with enough basic knowledge in that area, or pursue study elsewhere, as a foundation for discussion of these topics. With that introduction to dashboard technology it is now possible to talk about the applications for which the technology is intended.

DASHBOARD APPLICATION SITUATION ANALYSIS

Under the heading of dashboard applications, there is a wealth of new control products, technologies, and buzzwords. As noted above, due to the evolution toward data communication standards, middleware and web services, it is now possible to develop products faster and easier. For that reason, and due to the market growth in energy services, technology products of many types are being developed and launched. This chapter cannot cover all of the various product types, but it can provide a good foundation for future discussion. An obvious first application to discuss is building automation systems (BAS) interface and related applications, which in the past consisted of a very small number of manufacturer developed offerings. The BAS industry now includes dozens of companies offering many types of software and hardware, including dashboard interfaces, and equally importantly those that offer web services and analytics technology to expand the power of these systems. Under the in-depth discussion of analytics, another important development is the creation of a new organization, Project Haystack, -which was discussed in the last chapter and has an entire chapter devoted to it later in the book.

Market demand and reduced technology development barriers to entry and data standards have created a perfect storm of opportunity for dashboard products. As a result, many different players, not just name brand BAS manufacturers, may introduce new advances in BAS interface. This becomes more common as the lines blur between traditional BAS manufacturers and heating, ventilation and air conditioning (HVAC) manufacturers. As these companies continue to merge, grow and expand their product families, it becomes less likely that the traditional BAS manufacturers will be technology leaders. A strong argument could be made that these companies have not been technology leaders for some time. The real leaders today are part of this new ecosystem of software/technology companies developing products including dashboard and analytics tools. In many ways standards are "the price of admission" to leverage off-the-shelf technology and web services in integrated system projects. This is particularly true in the 21$^{st}$ century.

During the first wave of web-based products, the energy and buildings industry was inundated by IT developers who thought that building systems were like email or any other web service. Today however, the energy systems industry is much more IT savvy and the developers are coming from the energy and buildings world. This means visualization tools and other web based products are much better conceived and deployed than before. The stage is now set to go beyond seeing the dashboard as a BAS interface only, but seeing it an integration tool for the applications described in Chapter 1 as well as building commissioning, facility management including maintenance and operations, video surveillance, electronic access control, digital signage and much more. Dashboards are valuable tools to simplify interface and save time by unifying daily operations tasks associated with monitoring these systems. They become even more valuable when the act of interfacing makes it possible to achieve system integration. An example would be integrating fire and security with HVAC control for energy management. Since integration requires data and system level interface, it goes beyond dashboard monitoring. True integration will be more relevant as we discuss analytics. However, dashboards can be used for troubleshooting and diagnostics to ensure that integration sequences are operating properly.

The dashboard becomes a tool that brings together information so that mangers can make effective decisions. It is a portal that allows the manager to get access to real-time information about building operations, and to link to other systems, like building automation or access security, for interrogation and update. It can also provide access to other websites, such as the local utility site, or to internal software like the budget. This makes it possible to access data from the analytics tool and view current energy consumption and costing data from the local utility and retail energy providers. The dashboard can also report on sustainability metrics and provide interface to onsite generation from renewable energy or cogeneration.

This opportunity for dashboards is not without challenges however, and there are risks that must be managed. Cyber security is an ongoing concern, but the benefit of combining automation and information systems is that integrators can leverage existing firewalls and other security measures. In early 2012 an investigative report was published in the *Washington Post* about the "middleware" technology that was installed at sensitive government facilities. The article implied that hackers, and others prone toward terrorist activities, could gain access to the computer networks in these facilities through the middleware systems. The reporter asserted that the password security for these systems was not robust enough, and that a hacker could gain access to a password that would open the system. There were several flaws in this argument, but nonetheless extensive work was done to enhance the integrity of the system for the future. There were also upgrades to these technologies to protect password security within the systems. Data reliability could be another concern, particularly if the control information is completely reliant upon the web. Improving middleware security from cyber threats also addresses data reliability issues, but adding storage at the building level and uploading data more frequently to cloud storage addresses most of these issues. Of course all such obstacles can be overcome with these and other strategies as they have in other industries like online banking.

In light of the previous discussion about cyber threats, there is value in briefly addressing the topic of data communications and cyber security. Managers should engage experts to develop approaches and take precautions in using these systems, as they would with any IT technology. Anyone living in the 21st century is well aware of the threats that information based computer systems may face from unsavory characters. This is a major focus for the smart grid interoperability panel (SGIP). SGIP started life as an entity operated by the National Institute of Standards and Technology of the Department of Commerce, under a direct mandate from Congress in the Energy Independence and Security Act. Its charter was to identify standards for interoperability, both data communications and functionality, between the grid and buildings as well as all other electricity users. The reason SGIP is relevant here is that it has focused very specifically on cyber security. Through the SGIP's work the bar has been raised for internet and data communication security between any piece of equipment in a building and the grid.

Without question, integrators and astute managers will succeed in creating cost effective building environments by synthesizing industry information to leverage technology and optimize system control and facility management. Dashboards are the windows that will make information about those efforts easily accessible for a wide range of people and in user friendly formats.

# Chapter 4

# Introduction to Analytics

Thus far in this book the goal has been to frame the dramatic market opportunity that exists for energy and analytics. In Chapter 4, John Petze, Partner in SkyFoundry, LLC, co-founder of Project Haystack and author of the foreword to this book, shares a high-level view of what the technology offers. Mr. Petze is both a serial entrepreneur and a serial technology visionary, and initially developed this content for an industry technical paper. The content was then expanded for inclusion as a chapter in Barney Capehart's book, *Automated Diagnostics and Analytics for Buildings*, also published by The Fairmont Press. This very informative introduction to the emerging technology and business practice being deployed with analytics should be valuable to the reader. Of equal importance is that Mr. Petze also dispels some of the confusion regarding analytics and helps to highlight the differences between some technology-driven solutions that exist in the marketplace.

*Analytics, Alarms, Analysis, Fault Detection and Diagnostics*
Making Sense of the Data-oriented Tools Available to Facility Managers
*John Petze, Partner, SkyFoundry, LLC*

Using data from smart devices such as building automation systems, smart meters, smart sensors, etc., is one of the hottest topics in the industry. The goal is to use these data to better understand the operation of building systems and identify ways to improve that operation.

Facility managers are confronted with a wide range of products and features that can contribute to improved facility performance. From alarms to fault detection and diagnosis (FD&D) and analysis tools, to automated analytics, each has their place and offers specific capabilities and benefits. Systems integrators and owners, however, are often faced with comparing "apples and oranges" as they try to evaluate the different tools.

Comparing the technologies with sets of criteria can help facility managers better understand the roles, capabilities and benefits of these tools so they can assess the best fit for their needs. This chapter explores the range of tools from the perspective of:

- Time and location of implementation (when they are defined and where they run)

- Data scope—the range of data items being analyzed

- Time range of analysis

- Expressiveness of the tool—the tools available to describe the issue to be detected

## ALARMS

Alarms are one of the fundamental tools that have been available in BAS systems since the early 1980s and remain an important tool. Often, when first introduced to advanced analytic tools, people compare them with alarms. After all, doesn't an alarm programmed in a BAS tell the operator something is wrong? On a very basic level there is a similarity, but if the reader looks a little bit deeper they see that there are fundamental differences between alarms and more advanced analytic tools.

First of all, alarms require an understanding of what you wanted to look for at the time you programmed the system. In other words, you knew exactly what you wanted to look for and took the time to program that specific alarm definition into the system. This is fine for simple issues like a temperature going outside of a limit. However, there are many inter-relationships between equipment systems that may not be known at the time the control system is installed and commissioned.

One of the great benefits of analytics is that it enables you to find patterns and issues you weren't aware of at the outset of a project—providing results that show how your building systems are really operating vs. how you thought they were operating.

Given that alarms require you to define the specific condition ahead of time, *time of implementation* is typically during the initial programming of the control system. This requirement fits a wide range of conditions that we want to identify in our control systems, but is also a limiting factor.

## Data Scope

Alarms usually evaluate a sensor value vs. a limit. They may also include a time delay—i.e., the condition must be true for 5 minutes before an alarm is generated. Alarms are most often associated with a specific point. For example, one of the most common approaches is to set alarm limits for each individual point when it is configured. The data scope of alarms is also typically limited to the data in the local controller or other devices within the control system. Alarms do not typically evaluate enterprise data or data from other external sources.

## Time Range

Alarms are typically evaluated "now." By this we mean the real time condition of the sensor vs. the alarm limit. This is a key point—very different techniques are needed to look back over months or years of data to identify conditions patterns and correlations.

## Describing What Matters

The next difference to consider is the flexibility of expressing what you want to find. Alarms don't typically allow for sophisticated logic that interrelates multiple data items, conditions, data sources, etc. For example, an alarm definition might be: "Is the value of the room temp sensor above 76 degrees F right now?" An analytic evaluation on the other hand might be: "show me all the times when any room temperature was above 76 degrees in the last year for more than 5 minutes at a time during occupied hours, and total the number of hours by site."

## Processing Location

Adding new alarms typically means modifying control logic or parameters in controllers. This means you need to have access rights to modify the controller logic to change or create alarms. This can be very limiting if you just want to "find things" in the data or are trying to analyze data from a system installed and managed by others.

The need to "reach into the controller" makes alarms "expensive" when trying to use them as an analysis tool. For example, could we justify reprogramming controllers in 500 remote sites because we have an idea of a data relationship we want to look for? Most likely this would be cost prohibitive. We might also ask whether the "analyst" should even have access rights to the configuration tools in an automation system. The important point is that there is significant "friction" involved in using basic alarm techniques for anything beyond limit-based relationships of individual points.

## FD&D—FAULT DETECTION & DIAGNOSIS

FD&D techniques are typically equipment centric and characterized by pre-defined rules that are based on an engineering model of a piece of equipment—for example, FD&D rules for a type of packaged AHU.

### Time of Implementation

There are two "implementation time" components to consider with FD&D. Generally FD&D requires that an engineering model of the equipment be developed beforehand. In this respect they require significant pre-knowledge of the system. As such, FD&D rules are often not flexible for use on custom, built-up central systems, etc. The fact that no two buildings are alike can further limit where FD&D techniques can be applied.

Because of the dependence on predefined equipment models FD&D is typically not a good fit for *ad hoc* analysis—e.g., "I have this idea about a behavior I want to detect." In addition, FD&D rules can often be developed only by the software/service provider. The rules are "part of the product" versus being programmable on a project-specific basis.

### Processing Location

FD&D solutions are typically applied as a separate software application that pulls data from the BAS system. The software may be installed locally or hosted in the cloud. Some FD&D solutions can be programmed into BAS controllers. In this case they require "touching" the control system. As previously discussed, this can be a limiting factor in their application.

### Data Scope

FD&D rules are typically focused on the predefined points associated with a known piece of equipment. They may include data such as weather, but do not typically encompass external data, like age of building, historical energy consumption, type of facility, square footage, type of equipment, etc. or provide the ability to roll up and correlate data from hundreds of pieces of equipment.

## Time Range

FD&D rules typically look at real time conditions, but some have the ability to look at data from a sliding window of time—such as the last hour or day of operational data. *Ad hoc* analysis of random time periods (i.e., last August vs. this June) may not be available.

## Analysis Tools

Most often discussed in relation to energy meter data, analysis tools provide an experienced user with the ability to look at data with a range of charting tools and "slice and dice" that data with a range of tools to identify peaks, and other anomalies. Analysis tools typically include the ability to perform normalization against weather, building size, and other factors, etc.

The most significant characteristic of analysis tools is that they require a knowledgeable user to be sitting in front of a screen to interpret charts and graphs that identify the important issues—in other words, "wetware" is a key part of the issue identification process.

## Data Scope

Most commercially available analysis tools focus on a specific type of data and application, for example, energy meter data. They integrate weather data (degree days as a minimum), occupancy schedules and building size, but do not integrate the full set of equipment data such as temperatures, pressures, speeds or rate of operation, equipment status, etc.

## Time Range

From a time perspective, analysis tools provide the ability to analyze across a wide time range. As for "real time" data they can typically handle data "up to the last reading"—often a 15-minute sample—but are much less likely to be connected to a data feed that updates values every minute or second. They also support batch loads of historical data from meters, utility sources etc.

## Processing Location

Analysis tools can be applied on top of existing systems as long as the data are available in some open format. They do not need to be part of the initial installation and typically do not require any changes to BAS programming. Analysis software can be hosted in the cloud or installed on-premise.

**Analytics**

In many ways, analytics can be thought of as a superset of the other categories we have described. For example, analytics can be applied to "real-time" alarming situations and offer the ability to define more sophisticated alarm conditions to create "enhanced alarming."

FD&D rules that diagnose equipment performance issues are a type of analytics as well. While most FD&D solutions employ pre-written rules based on known models of equipment, programmable analytic tools enable experienced engineers to implement rules based on their knowledge—they are not limited to rules defined by the software provider.

In comparison to the other technologies analytics have the following characteristics:

*Automated Processing*

An analytics engine continuously processes data to look for the issues that an experienced engineer would normally look for manually. This ability to automatically process rules to identify important patterns and correlations is the hallmark of modern analytics solutions.

*Time of Implementation*

Analytic solutions can be implemented anytime, during initial installation or years after. They do not require reaching back into the control system to make programming changes for analysis. They do of course require that data be accessible (we will talk more about data availability in a moment).

*Expressiveness—Flexibility to Define Rules for*
*Conditions to be Detected*

While a typical alarm might evaluate a single item against a limit at a single point in time—analytic rules crunch through large volumes of time-series historical data to find patterns that are difficult or impossible to see when looking only at real-time data.

For example, while an alarm might tell us our building is above a specific kW limit right now, analytics tells us things like how many hours in the last 6 months we exceeded the electrical demand target. How long were each of those periods of time, what time of the day did they occur, and how were those events related to the operation of specific equipment systems, the weather or building usage patterns?

Analytic rule languages should enable sophisticated data transformations beyond limit checks. Examples include: Rollups across time periods, calculation of max, min, average, interpolation across missing data entries, linear regression, correlation of data sets to find patterns such as intersections (or lack thereof), etc.

One of the key characteristics of analytics is that they expose things you were not necessarily looking for, or even knew to look for. Analytic data presentations expose data relationships and correlations even without writing rules. And systems that offer user programmability enable new rules to be implemented as findings illuminate actual operating characteristics, and new priorities emerge due to changing energy costs, operating requirements or building usage patterns. In fact, the successful application of analytics is a journey with one discovery providing insight for additional analytic rules. More on this later.

*A Wide Data Scope*

Analytics enable multiple data sets from different sources, in different formats and with different time sampling frequencies. They are not limited to data within a controller or a control system. Examples might include:

- Energy, weather, and control system data

- Size of facility (sq ft/sq m)

- Age of building

- Type of building and systems (packaged HVAC vs. built-up central systems)

- Equipment brand

- Service company

In many cases the analytics process starts with data available without establishing live connections to control systems, meters of other devices. More on this topic shortly.

Analytic tools can be applied on top of existing systems as long as the data are available in some open format. They do not require changes to the control system and do not need to be part of the initial installation. This is a key benefit, as it allows analysts to apply rules to the data without disturbing underlying systems.

Analytic software can be hosted in the cloud or installed on-premise. Each approach offers tradeoffs. For example, hosted solutions allow the software to be managed centrally, but require that the customer accept an external connection to their network for continuous access to data.

## SOME EXAMPLES HELP TO HIGHLIGHT THE DISTINCTIONS

### An Alarm
Detect zone temperatures above 76 deg F when occupied.

### An Analytic Rule
Look at signature of data associated with all sensors to indentify "broken" sensors or sensors out of calibration. See Figure 4-1.

### Alarm
Detect kW above a specified limit in real time. See Figure 4-2.

### Analytic Rule
Identify periods of time when demand is above a specified kW limit (see Figure 4-3), calculate cost impact, make reports available showing duration and even cost across any selected time frame, and provide continuous real time processing of the rule as new data are received.

### An Analysis
Generate a graph of energy consumption across a specific time.

### Analytics
Automatically correlate equipment operating status with energy consumption across a specific period of time. See Figure 4-4.

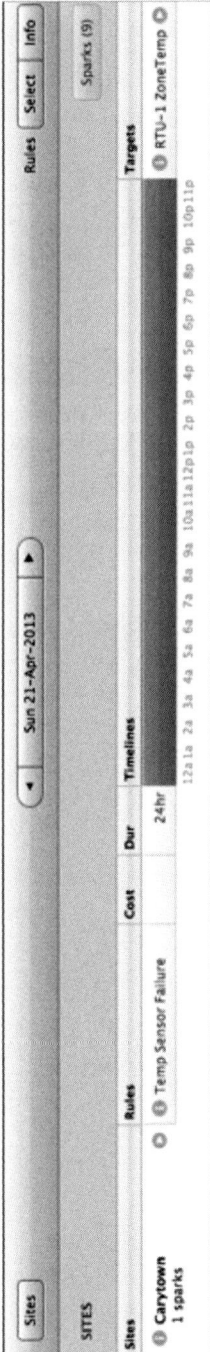

Figure 4-1

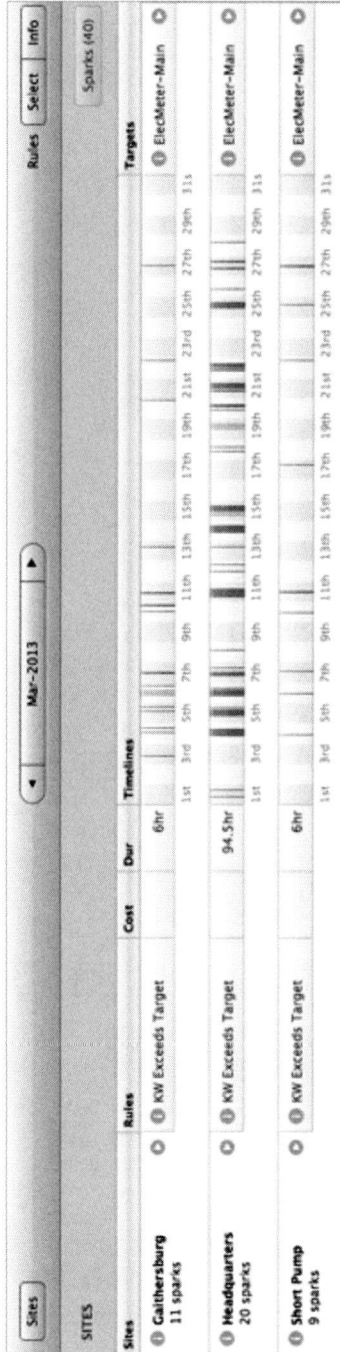

Figure 4-2

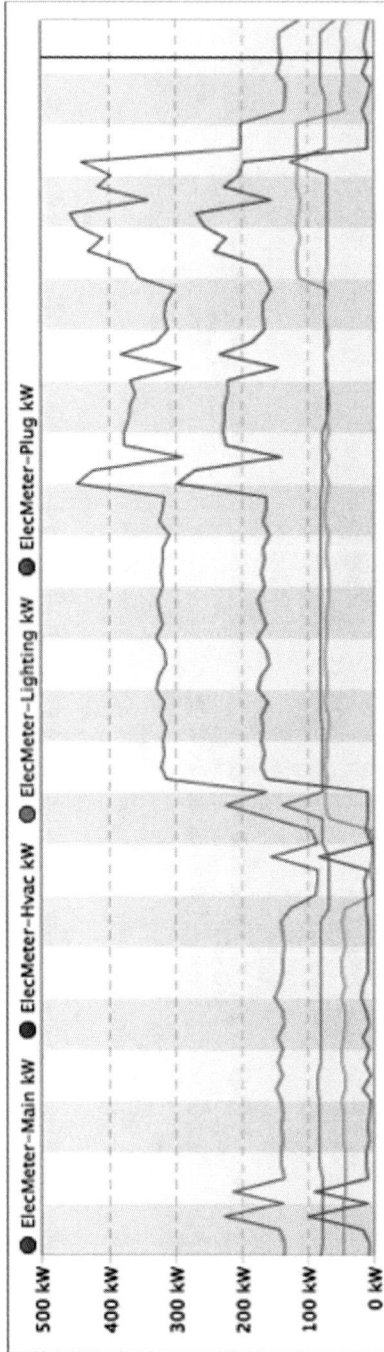

Figure 4-3

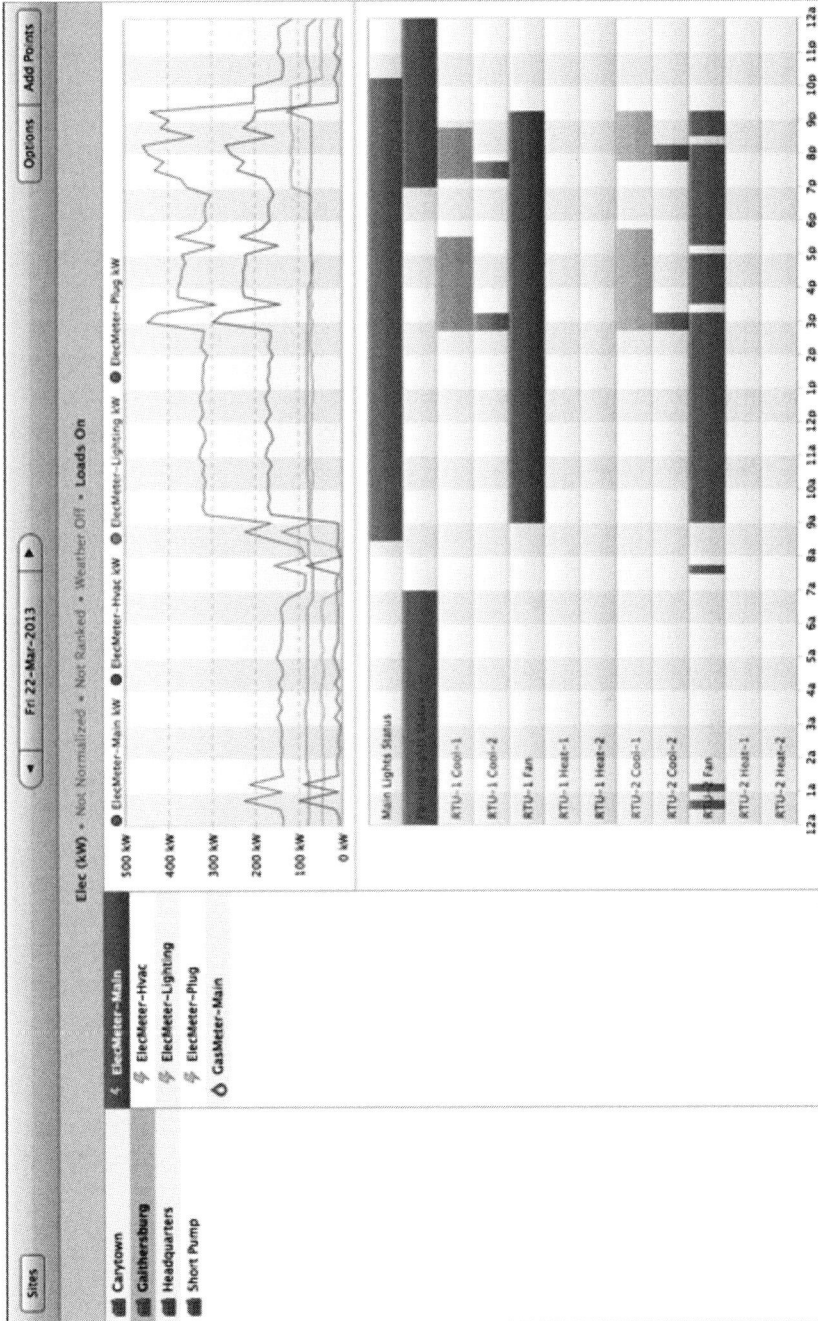

Figure 4-4

CAN ANALYTICS EFFECT CONTROL ACTIONS?

Once analytics detects a pattern of interest can the system act on it? The answer is yes, analytics can be used to issue commands to control systems, but it's important not to overestimate the applicability of this feature. Many issues found using analytics will not lend themselves to being corrected with a simple "command" to the control system. Two quick examples are illustrative:

1. Errors in control strategies. If an analytic rule detects conditions resulting from defects in a control sequence, the logic will need to be modified. An example might be simultaneous heating and cooling. While you could command the heating or cooling off when detected, the correction of the problem will require actual changes to the control sequence.

2. Physical equipment issues. If a damper linkage is broken or stuck, or a valve won't fully close, or a sensor is out of calibration or not reading correctly, there is no command to "fix" these issues.

This doesn't mean there is no use case for automated response to analytic results. Two examples of an automated response to analytic findings are:

1. Automatic generation of more intelligent work orders by the integration of analytics and CMMS tools.

2. Automated demand response. Demand response actions can be executed in response to energy use patterns detected (or predicted) using analytics. This provides more intelligence than simple limit-based demand response control.

DATA ACCESS—A KEY REQUIREMENT FOR
ALL DATA-ORIENTED TOOLS

All of these tools are dependent on being able to access relevant data. Because alarms are processed locally in the control system, data access is not an issue—the data "is there." In order to take advantage of the other tools, however, we need to assess the process to gain access to the data. A good way to start an assessment is by answering the following:

1.  What data do you have? Examples: Energy meter data, facility data (size, location, type, year of construction, etc.), equipment operation data such as on/off status, sensor data, etc.

2.  Where are the data located? Examples: BAS system, SQL database, utility company website, Excel spreadsheets, etc.

3.  What method will be used to access them? Examples: Live collection of data via BACnet™ or oBix™, Haystack™, Modbus™, etc., data download from utility website via xml (perhaps Green Button™ data), CSV file import, SQL queries processed on a daily, weekly, hourly or minute basis. The answers to these questions will vary dramatically based on characteristics of the specific project and customer's needs.

## ANALYTICS AS AN EXPLORATORY PROCESS

Analytics needs to be viewed as an exploratory process. Analytics show you how your buildings really operate, identifying where opportunities for savings exist, which assumptions are correct and which are not. Results from one rule or algorithm often identify behaviors that provide insight into other rules that should be implemented.

Because of this, the analytic process is best applied incrementally, driving value to the owner with each step. We have a tendency to think of any energy conservation measure as a big capital expense project. Unlike energy conservation measures that involve the installation of major capital equipment, you can start small with analytics and generate returns in a very short period of time—you don't have to "do it all" at once to get value from analytics. The results from initial analytics generate the savings to go deeper into your operational data. This is a great advantage to building owners, in that it's possible to start with a limited amount of data, and a low-risk, low-cost initial evaluation, to produce short-term financial results.

### An Example of a Shallow and Wide Approach

Basic interval energy data, combined with building occupancy, and weather data are a great place to start as shown below. Let's say the only data I can easily access are my interval meter data (kW demand), which are provided once per day by my utility company, and a list of occupancy schedule times in an Excel™ spreadsheet (see Figure 4-5). We can gain any valuable insight from that limited amount of data. With just that limited amount of data SkySpark can identify:

- Buildings starting early,

- Buildings running late,

- Buildings that operate continuously,

- Demand peaks that occur outside of occupied times,

- Peak load, annual and monthly, and short load durations. See Figure 4-6.

Here's how: Using conventional energy analysis tools I can easily identify my kW demand pattern for each day. By looking at it manually I can determine whether the pattern follows occupancy times.

If I have a large number of buildings this type of manual review would be prohibitive. With analytics I can automate the process and have the software find these patterns for me. A rule can look for a percentage change in kW demand at the transition to and from occupancy. The rule generates "notifications" whenever the demand does not change by the expected amount within a defined period of time around the occupancy transition. In this way the system automatically identifies when buildings run late, and start early. A weekly (or monthly) view shows how many times the issue has occurred and the total cost. There's no need to hunt through the data manually.

**Do They Have to be Live Data?**

Another common misconception is that you can't derive value from data unless they're live and continuously updating in real time. This simply isn't true. Live data are great, but by no means essential to getting started with analytics. You can get tremendous value from running analytics on a snapshot of historic data. One of the benefits of using snapshot data is that you can avoid the costs and delays associated with gaining IT approval for network access to live systems. A great example is an initial portfolio assessment, but you can also do deep equipment analytics with historic data.

It's worth digging into the topic of "real time" data a bit more. We have found that there is a significant variation in what people mean when they use the term real time data. A few definitions can help clarify the topic and enable a clear understanding of how analytic tools can be used with "real time" and historic data.

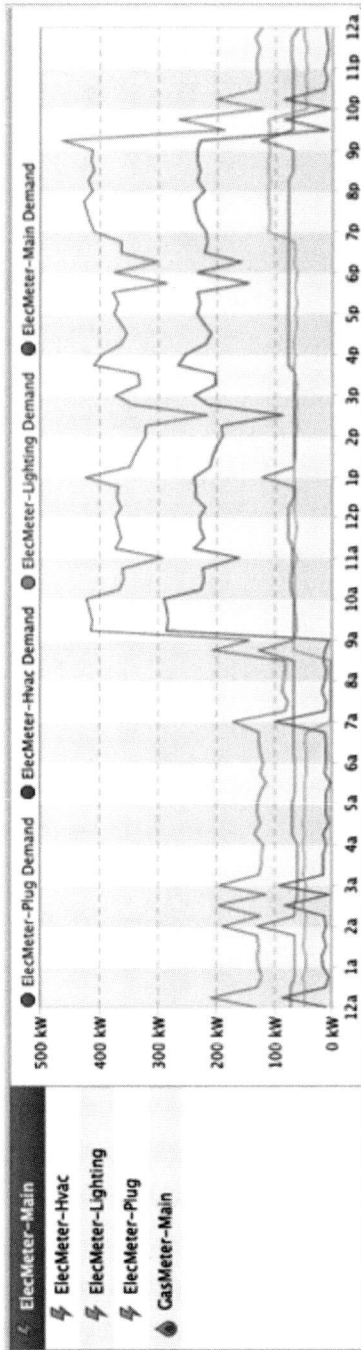

**Figure 4-5**

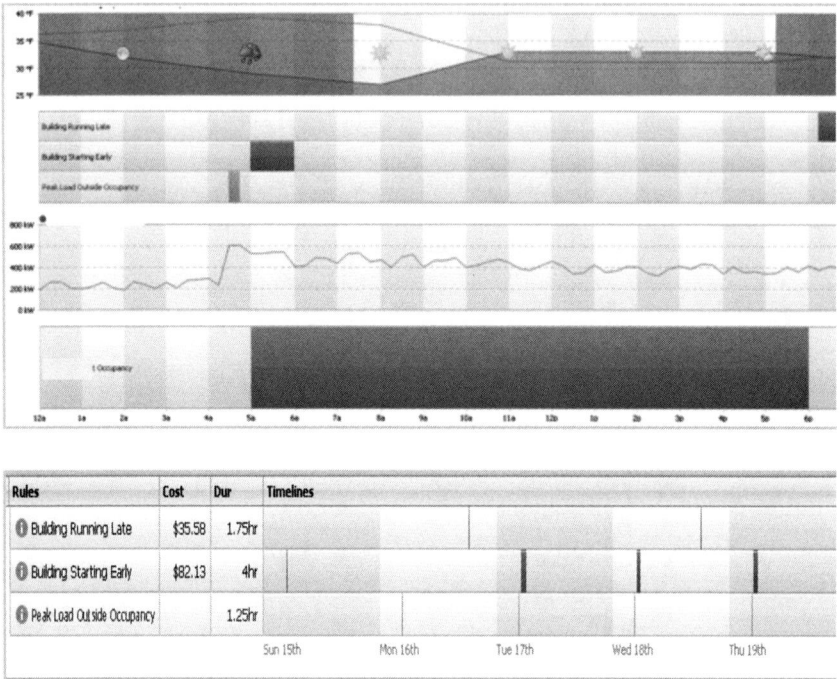

| Rules | Cost | Dur | Timelines | | | | |
|-------|------|-----|-----------|--|--|--|--|
| ⓘ Building Running Late | $35.58 | 1.75hr | | | | | |
| ⓘ Building Starting Early | $82.13 | 4hr | | | | | |
| ⓘ Peak Load Outside Occupancy | | 1.25hr | | | | | |
| | | | Sun 15th | Mon 16th | Tue 17th | Wed 18th | Thu 19th |

Figure 4-6

## A Definition

Real time means fast enough for the application. So if we are controlling a piece of industrial manufacturing equipment, that might mean we need a control loop response time of 10 milliseconds to meet the real time needs of the process. Many processes require even faster real time requirements—perhaps on the order of microseconds.

For a VAV box controlling temperature of a room, the control response time of the temperature control loop might be 30 seconds, a minute or longer. On that same VAV box, though, the control loop responsible for managing the airflow volume to maintain an airflow setpoint might have a response time requirement of 1 second.

At the other end of the spectrum, in energy metering applications real time often means 15-minute interval data. When we are looking at applying analytics to data there are a few facets of "time" to consider.

## Resolution of Data—The Sampling Rate

One aspect of "real time" we need to consider is the resolution of the sensor data. By this we mean how frequently a system senses (and records) data values. For example, many BAS controllers can store data values on the second. In this case the data-sampling rate would be once per second.

The sampling rate of sensor data has a major impact on the volume of data created. For example a history record of zone temperatures once per minute will generate 1,440 values per day. That same sensor recorded on a sampling rate of 1 second will generate 86,400 sensor values—sixty times as much data!

In most systems it's common that the control response capability and sampling rate are equal. For example, a controller capable of providing 1-second control response will often allow sensor values to be recorded once per second. One major limiting factor, however, is that most controllers will typically run out of storage space if you choose to record sensor values every second. So in reality they are often set to record values once per minute or longer, or need to upload the history data to a repository on a frequent basis.

## Freshness of Data

The next concept to consider when discussing real time applications is the "freshness" of data or frequency of the updates to the software application consuming the data. A common example of this would be the update frequency of sensor data on an operator screen such as a graphic representation of an air handler. The capability of a system to deliver updates to the user is affected by a variety of factors including:

- Bandwidth and speed of the communication network,

- Computing resources of the controller that can be dedicated to communicating data to the UI application,

- Efficiency of the software that presents that data to the user,

- Computing resources of the computer that is displaying the data.

Different systems vary considerably in this area and it's not uncommon for screen updates of 10 seconds to be considered "real time" updates. Going back to our definition of real time, the "application" here can be described as the ability (or need) of the human operator to read and respond to changing values he sees on the screen. Based on this definition a 10-second update frequency may be considered adequate.

The data source—the controller—may be operating its local control loops with a response time of 1 second or 100 milliseconds, but the fastest that new data will appear on the screen might be once every 10 seconds or longer. In some building automation systems it's not uncommon to see screen update times be much longer. That might be perfectly acceptable for the "application" of the human user (although most people find that a bit frustrating).

**Application of These Principles with Analytics**

When planning the application of an analytics solution, all of these factors need to be considered. For example, knowing the sampling rate of the data will enable us to understand the richness that analytic rules will be able to work with—think about the different level of insight provided by 15-minute-interval meter data versus monthly energy data. The sampling rate will also tell us about the volume of data we will need to store and manage. In many applications there will be different data sources, each with its own sampling rate. The analytics application must be designed to handle the challenge of aligning and analyzing data with different sampling rates.

Next there is the question of how frequently data can be pulled from the source. Here again different data sources may have different update frequencies. Data freshness is related to the application. The frequency with which you pull new data into the analytics database varies based on the needs of your application and limitations of your communication infrastructure. For example, you might have an application where you can only upload data from a site once per night at midnight (we've seen these restrictions placed on systems by IT departments). At that time, however, you might be pulling in data that has 1-second resolution. So in this case, the data freshness would be once every 24 hours, but the data sampling rate would be once per second. The sampling rate and the freshness are "decoupled." Then there is the question of the frequency of the rule processing. Ideally, rule processing is decoupled from the sampling rate and freshness of the data.

Next consider the user of the analytics application. Using the same example, we might have a situation where building operators want to see their analytic results (sparks) every morning. They could subscribe to a daily digest notification, which will provide a daily summary email. When they view their sparks they will be seeing all of the issues detected in the last 24 hours of data. Their "response time" would be on the order of a day (about the same time frame as the data freshness).

Too slow you say? It depends on the application and needs of the user—there is no one-size-fits-all answer. For example, we have seen some examples where the data sampling rate is once per minute, the data freshness is once per 15 minutes, and yet the update provided to the user (the building operator) is once per week or once per month. That's right, building operators are informed of sparks once per week! Why might an operator find this level of "freshness" in sparks acceptable or even desirable?

It could be because the types of issues they are looking for can't really be addressed more quickly. Perhaps they need to be planned into their facility maintenance schedule, which is set on a weekly basis. Or it could be because they have so many issues requiring attention that they simply can't do anything with issues presented on a more frequent basis.

It could also be because the issues they are tracking take time to form—the patterns appear over a period of time. Some patterns form in a minute, some take hours, some take days and some might take a season or a year to form. This is a key point where people often confuse analytics with alarms. For example, the pattern that represents a defective sensor might require 12 or 24 hours of operation to detect. It can't be detected at any specific second in time—it is detected by interpreting a pattern in data that appears after some period of time has elapsed.

All of this leads us to appreciate that there is yet another "decoupled" response loop involved in the use of analytics—that of the corrective action response time. The limiting factor here is typically the human systems that will respond to analytic results. An issue detected by analytics might not be able to be addressed until the next planned service call to the facility.

## A Fast Moving Field

None of the examples presented is meant to be absolute, rather they are offered to help systems integrators and facility managers gain an understanding of these tools, their requirements and potential benefits. With the rapid advances in data-oriented facility management tools, there is some overlap between them, and the lines blur as vendors advance their technology.

As we stated in the introduction, each of these tools plays an important role in achieving an efficient building. Analysis tools help us gain insight into operating characteristics, which then support automated analytics to provide continuous detection of important issues relating to equipment operation and energy use.

# Chapter 5

# Analytics for Energy Management in Buildings and on Campuses

In Chapter 1, the four essential applications for analytics were introduced. Energy was one of those applications for analytics, and has been highlighted as the focus for this book. Optimizing energy consumption is central to addressing the complex challenges facing business, government, institutions, the economy and the planet, and it is required to achieve the high standard of living that is sought by every nation in the world. These aspects of life on the planet are all reliant upon energy, and energy is also highly interdependent with the other three applications for analytics. Moreover it is essential to mention the old adage "one cannot manage what one cannot measure." It will be cited more than once in this book, and this statement is particularly true for energy. The emphasis in this book will also turn heavily to electricity, because it is the highest quality form of energy, but more importantly because the electricity market creates unparalleled challenges and opportunities to leverage analytics for significant financial, operational and mission (work done in the building) benefits. The preceding chapters have provided clear justification for engaging in a robust and persistent analytics program. This one will reinforce that justification with a focus on energy.

This chapter and the two that follow will begin to provide more specific content about deploying analytics for targeted applications and opportunities. In the author's mind, electricity holds some of the most significant opportunity of any form of energy, but the interruption of electricity flow also presents a significant risk. The number-one reason for any manager to engage in energy and analytics is to keep these two dynamics, opportunity and risk, in balance. The opportunity will be explored further here. At the same time, electricity resiliency, already defined as the reliable and continuous flow of power, is essential to this topic of analytics for building owners and managers. Some may say that there is nothing a building manager can do to support energy resiliency, but this author would disagree. This book is about the full spectrum of opportunity that analytics can offer to aide building owners/operators and Certified Energy Managers (CEMs) in maximizing the use and cost of energy, as well as strategies that can improve the environments created within the building. The resiliency of energy, especially electricity, to support the mission, or the work that is being done in that facility, offers the opportunity to forge a true 21st century energy management program.

Management professionals, in the building community today, must oversee enterprise infrastructure to address a wide range of short- and long-term objectives. Many of these objectives are in some amount of conflict; such as creating the optimal balance of occupant comfort with energy cost reduction. Some might ask, how does energy and analytics technology fit an overall program for building operations and energy management? Lawrence Berkeley National Laboratories (LBNL) is among the national labs that has established leadership in this area. LBNL's Electricity Markets and Policy Group has conducted technical, economic, and policy analysis of energy topics centered on the U.S. electricity sector. Much of this research is intended to aide in public and private decision-making, by managers, on issues related to energy efficiency and demand response, renewable energy, electricity resource and transmission planning, and electricity reliability. One report is cited here, but the reader is encouraged to research the wealth of information available from LBNL, Pacific Northwest National Labs, the Office of Energy Efficiency and Renewable Energy within the U.S. Department of Energy (DOE) as well as state and local agencies and resources. The bottom line is summarized in a 2008 DOE report called Energy Efficiency Trends in Residential and Commercial Buildings http://apps1.eere.energy.gov/buildings/publications/pdfs/corporate/bt_stateindustry.pdf. That report points out that buildings account for 40% of all energy used annually in the United States. By energy type, buildings account for 72% of U.S. electricity use and 36% of U.S. natural gas use. Given the significance of building energy use, it should come as no surprise why measurement, and active management, are so important, and why there is a growing market for energy analytics.

## HISTORY AND PROCESS

History is a big topic and cannot be covered in detail, but there are some important topics that must be touched upon. First and foremost, management of energy requires measurement of how energy is consumed. In fact the overarching term that is used to refer to the discipline is "measurement and verification" (M&V). Many people think of M&V as the process to evaluate energy investments, and it is, but before the investments it is critical to develop a thorough understanding of energy consumption baseline measurement. It is only reasonable to assume that most readers are well versed in these topics, however the author will ask your indulgence to provide some perspective for those who may not be knowledgeable in this area. Equally important, this section will be extended somewhat because it is also logical to describe the process of energy analysis, and how some of these historical evolutions added to what has become a very robust discipline. The formation of this discipline that we now call *"Energy Analytics"* is based on work done by the DOE, Association of Energy Engineers and many other thought leaders in the energy management world, and it begins with a simple mantra: You must baseline. Baseline means capturing a history of energy consumption including electricity, natural gas, oil, gasoline, propane, etc., and summarizing that history in a tabulated energy profile reporting on monthly and annual utility cost for at least two years. The best place to start is with the "baseline," which could be called measuring the building energy intensity. In fact some have referred to the baseline as an energy intensity index. Developing the baseline typically begins with gathering utility bills for a building, and in the beginning this was done by contacting each utility, requesting copies of the bills and then summarizing the information with pencil, paper and calculator. Today most utilities make this information available on line. Of critical importance, especially for managers of multiple facilities in different geographies, utility territories, etc. is that it is important to "normalize" the baseline.

Normalizing means that all of the energy consumption is translated to a standard form of unit, typically a British thermal unit (Btu). Most readers will know that 1 Btu represents the amount of heat that is created by one match burning, and that there are Btu equivalents for all units of energy—kWh, Therm, etc. Normalizing will also include a way to compare facilities, so it usually includes a square footage calculation and a weather calculation. The universal standard for weather is a degree day; heating degree days count each degree of average outside temperature during 24 hours that is below 65 degrees Fahrenheit, and cooling degree days count each degree of average outside temperature for 24 hours that is above 65 degrees Fahrenheit. To calculate the heating degree days for a particular day, find the day's average temperature by adding the day's high and low temperatures and dividing by two. If the number is above 65, there are no heating degree days that day. If the number is less than 65, subtract it from 65 to find the number of heating degree days. For example, if the day's high temperature is 60 and the low is 40, the average temperature is 50 degrees. Sixty-five minus 50 is 15 heating degree days. Cooling degree days are also based on the day's average minus 65. For example, if the day's high is 90 and the day's low is 70, the day's average is 80. Eighty minus 65 is 15 cooling degree days. Degree day information is available from a number of sources including the National Oceanic and Atmospheric Administration (NOAA) www.noaa.gov. So step one in analysis is to create a baseline of annual energy consumption by month and year that is normalized to arrive at Btus per square foot per degree day per year. This process is sometimes called developing a "building energy profile."

The next step is to use this information to understand how the building is performing, and a good way to do that is to compare it with other buildings. The holy grail for building energy performance data is the Energy Star for Buildings and Plants (Energy Star) program of the U.S. Environmental Protection Agency (EPA), and its ENERGY STAR Portfolio Manager® software. Portfolio Manager is an online tool you can use to measure and track energy and water consumption, as well as greenhouse gas emissions. Sustainability was mentioned in Chapter 1, because carbon is an organizational performance metric that has been established for many buildings. Analytics can be very valuable in tracking sustainability metrics, but it will not be discussed a great deal in this book. Most readers are already aware that greenhouse gas emissions and energy consumption are inextricably linked. This was underscored by the EPA ruling targeting greenhouse gas emissions from power plants, which is discussed more in Chapter 7 and emphasizes the alignment of these topics. The EPA discusses this in its report "Regulatory Impact Analysis for the Proposed Carbon Pollution Guidelines for Existing Power Plants and Emission Standards for Modified and Reconstructed Power Plants," June 2014. In that report the EPA states that carbon dioxide ($CO_2$) is the primary greenhouse gas pollutant, and it accounts for 84% of U.S. greenhouse gas emissions. According to the report, the single largest emitters of greenhouse gas, primarily in the form of $CO_2$, are fossil fuel fired electric generating units or power plants. At the building and campus level, it seems that most organizations have developed sustainability plans to address "carbon" and to add this factor to the analysis that is conducted for the building. Again sustainability will not be covered in great deal in this book, however the energy and analytics tools discussed here, such as Portfolio Manager, are often used to measure and track carbon emissions as well as energy consumption.

Of particular significance in tracking, measuring and understanding how buildings consume energy is being able to have a frame of reference for how the building performs. Clearly there is value in developing a building energy profile and comparing the building to itself over time, but it is even more valuable to compare the building baseline to other similar buildings. The process of comparing buildings to one another based on performance is called "benchmarking." Again the EPA Energy Star program offers significant help with this task, because it has published a database of such information including consumption profiles by building type. This allows building owners to benchmark their buildings against similar buildings, and truly understand how they are performing. Most readers will also know that it is not useful to compare a hospital to a church, because the equipment installed, consumption patterns and hours of operation are completely different. For this reason Energy Star provides information to allow energy managers to compare similar buildings, normalizing for square footage and weather.

This section is about the history and process of energy analysis, and how energy and analytics are revolutionizing this discipline, but it can't explain every approach in technical detail. Therefore the focus will be on what energy analysis entails and why, rather than on how specific software developers, etc. complete the task. This author was a corporate energy manager for a retail chain with 20 million square feet of buildings in 23 states, purchasing electricity and natural gas from dozens of utilities. At that time, VisiCalc was the tool of choice for spreadsheet/paper and pencil analysis, but the predominant tool today is, of course, Microsoft Excel™. The point is that tracking energy in this way is a manual process, and the size/complexity of the dataset make it difficult to do this without errors. A few decades ago some very forward thinking professionals launched what became known as the "Enterprise Energy Management" (EEM) industry. These products performed all of the history and normalization functions of paper and pencil or spreadsheet, but purportedly with many fewer errors. Spreadsheets are wonderful, but many, many times they become works in progress and undergo continuous changes that open up the potential for errors and flawed reports. EEM tools on the other hand were developed by software engineers who conducted bug correction, with the intent of eliminating errors. Equally important was that these tools could calculate energy performance metrics on an ongoing basis and present reports that included graphic visualization charts depicting energy profiles. This visualization allows managers to see how energy profiles change by time of year or season, by differing periods when occupancy changes (think K-12 school or university) and based on various other factors. Another development that was introduced during the EEM era was software that maintained history and energy profiles, but these tools also made it possible to enter utility rate structures as well. Including the rate structure in the software made it easier for the analysis to determine several important things. First, comparing billing data to the rate structure makes it possible to verify that the building is on the right rate. It is not uncommon for some anomalies in energy consumption, particularly with electricity, to cause a building to be put on a rate structure, in error, that is more costly. Whether the building belongs on that rate structure should be the subject of ongoing analysis. At the same time, efficiency projects and active energy management programs may also qualify a building to be put on a different rate that is more advantageous. So re-evaluating rates is important, but it is time consuming, and software analysis can be very valuable. A very sophisticated tool may even be able to look for billing errors that can also be costly, and are rarely picked up by the energy consumer. From a historical perspective, the EEM market gained a certain amount of traction, but initially these were stand-alone packages, and required dedicated staff time to operate and use. They were particularly appealing to large users with campuses of buildings or multiple buildings in different geographies, and were of particular value to those who had buildings in many utility service territories. To a great degree, EEM has evolved to become part of the energy and analytics tools that are the focus of this book.

Energy measurement and profiles based upon billing data are very useful in getting a high level understanding of the buildings' performance. This brings even greater value when benchmarking is done. As discussed, baselines can be developed for any commodity (electric, natural gas, propane, oil, water, sewer, steam, refuse, recycling, telecom, etc.) and any number of accounts, meters, bills, and bill details. Baselines allow the manager to view each account's cost and consumption history in graphical or tabular format, and easily compare fiscal year, monthly, or annual summaries. Armed with this baseline data it is then possible to benchmark any building or account against others with information such as that available from Energy Star.

Benchmarking is extremely valuable, but for electricity more information is required. Electricity is a very unique commodity that requires more than a billing-based analysis. That is because of the impact that electric kilowatt (kW) demand fees introduce to the billing process. Even more importantly, kW demand is integral to how utilities create rate structures and to how the owner should manage electric use. The term that has become common when discussing kW demand is "interval data." Throughout this book the topic of 21st century energy, most importantly electricity, and systems will be discussed, and no topic is more important to that discussion than interval data. Interval data in this book will be concerned with kW demand. It is possible to capture interval data for kilowatt hours as well, but the opportunity discussed here is more focused on demand. Tracking interval data for kW is essential to optimizing energy use through demand management. Demand management can include demand limiting, as well as optimizing cash flow through demand response. Further demand management is essential to sizing for every type of distributed generation from solar photovoltaics to combined heat and power. The term "interval data" most likely comes from the fact that kW demand is measured on an "interval of time," typically 15 or 30 minutes. At some point in the history of electric regulatory policy, utilities were able to convince state commissions that energy charges alone ($0.00 per kilowatt hour consumed) were not sufficient to maintain the electricity system. Maintaining that system, or grid, requires utilities to cover all of the costs associated with operating existing electricity infrastructure, but it also requires that utilities cover the capital cost of building new infrastructure (generation, transmission and distribution systems). One of the ways that utilities are able to support that capital cost is to assess demand charges. This creates challenges for the building owner because electricity demand varies over time, and reality for the electricity business is that utilities must be able to meet that changing demand. For this reason, utilities must build enough power plants, transmission and distribution systems to meet the peak demand. An alarming fact is that as much as 25% of the U.S., and even worldwide, electricity infrastructure is in place to meet the demand for electricity for about 100 hours per year. Peak demand occurs for relatively short periods of time over a handful of days each year, typically driven by air conditioning demand on the hottest summer days. So the baseline electric analysis for buildings must include the energy consumption charge in kilowatt hours and the demand charge in kilowatts. It is critical to also add a record of interval demand data over the year as well. Of particular interest to the consumer is that the utility bill only contains one peak demand reading, the highest reading for any interval during the billing month. To get all of the interval kW data over an entire year, the owner must request special metering equipment that allows them to capture and record every interval during that billing month, so that they can analyze their demand peaks. The patterns (or profiles) of energy usage contained within this interval energy data are great for discovering how and when the building consumes energy, and can be very informative in identifying where savings can be achieved. The fine-grained detail of interval data (such as 15-minute or 30-minute data) is key—daily, weekly or monthly data don't carry nearly as much information about how energy is being used. EEM and most modern analytics programs will evaluate interval data.

Initially EEM functionality resided in stand-alone software, which was typically "shrink-wrapped," "licensed" and sold to the consumer. The database would therefore reside on the local computer. Today it is more common for this functionality and the data to reside on services in a cloud computing environment, and they often sold via software as a service (SaaS) rather than shrink-wrapped. Today's systems therefore may incorporate EEM, as an added module, and combine it with analytics for other applications. These cloud based analytics systems may be built around any number of applications, such as dashboards for visualization. Again the focus here is energy, and the author firmly believes functionality should not be sacrificed because it has been incorporated into other types of software. The goal here is simply to restate the importance of establishing requirements and carefully evaluating functionality before choosing a solution. The growth of the analytics market has resulted in the integration of this functionality into many offerings that go to market in a variety of ways. The balance of this book will cover this content from a variety of perspectives. Approaches to analytics tools will be discussed further in Section II. The topics of baselines, benchmarks and interval data will not be elaborated further here, but they are essential topics to energy and analytics. Readers who need to understand these topics further are encouraged to refer to the reports, etc. highlighted here and to explore training and further information from the Association of Energy Engineers, as well as state and federal agencies.

## Bringing the Building into Focus

The value of developing baselines and benchmarks for a building, campus or entire large portfolio of properties is significant. However, it becomes evident quickly that a more detailed understanding of the building, and the work being done in that building, is critical to making meaningful change. This is why the author discussed the four applications for analytics in Chapter 1, and emphasized the importance of an overall energy management program. Bringing the building into focus means having a team of content experts conduct a thorough and detailed building inventory and analysis. In some circles this might be called an "investment grade audit," but the goal is not to jump to identifying improvement measures, rather to develop a granular understanding of what makes up the energy baseline. There have been numerous articles and texts written on the topic of conducting a detailed building analysis. The emphasis here will not be on the steps to complete that analysis, but on the importance of this work and what to do with the information. At the end of the day, bringing the building into focus is about conducting a "forensic energy analysis." This process evaluates the energy consumption history and the interval data for electricity to identify what "equipment" and "occupancy/operational patterns" are causing the energy profile to look the way it does.

This building analysis must consider three important topics: 1) building type and characteristics, 2) baseline utility data and alignment with mechanical/electric equipment loads and 3) whether a building automation system (BAS) and metering technology are installed. The building type and energy history have been discussed already, but in summary these are topics that should be evaluated:

- Type of building (office, retail, university, hospital, etc.)

- Number of buildings and their locations
- Building size(s) in square feet
- Hours of operation
- Critical areas of the facility (data centers, operating rooms, laboratories, etc.)
- Sensitivity to environmental changes (HVAC and lighting)

Once these key factors about a building are understood, the equipment inventory is the next essential step in this process, and it should be conducted by Certified Energy Managers (CEM) who understand the use of that building. CEM is the term being used here because CEMs are usually highly diverse professionals with content expertise in engineering, equipment, systems, energy management and applications. The team may be made up of CEMs, mechanical or electrical professional engineers and/or facility management professionals. The CEM should develop a load profile that is validated against actual electrical usage data from utility bills. Elements of this analysis include:

- Electric Utility Accounts
  — Number of meters/accounts
  — Are interval meters installed? If so, what interval (hourly, 15-minute, etc.)
  — Utility rate program (standard commercial, interruptible, time-of-use, etc.)

- Load Profile Align with Energy Baseline
  — Annual consumption (kWh)
  — Summer and winter peak demand (kW) and what time of day does the peak typically occur?
  — Primary loads that impact the load profile (HVAC, lighting, process, etc.)

- Equipment
  — Significant electrical loads (process loads, large equipment, etc.)
  — What operations program could be implemented with this equipment/load?
  — Is there any on-site backup generation?

— How is cooling and heating supplied, and by what energy source?

In the end it is equally critical to understand both the equipment and the application (building function/use) of the building to do this job well. A thorough inventory of mechanical and electrical equipment will allow the team to begin to understand what makes up the kW peaks in the interval data. In most cases it will then take a content expert on the application (i.e., hospital, school, etc.) to analyze how the building is operated. There are still buildings that are operated inappropriately because staff are trying to avoid conflict, keep the maintenance office phone line from ringing, etc. There are also buildings that are operated inappropriately because of design errors, installation errors and changes in the building that have not been mirrored by changes in equipment configuration, sizing, operations, etc. Whether caused by one of these conditions, or any number of other issues, the forensic analysis process is the same. Using the inventory, the CEM can then look at electrical loads, occupancy and operations patterns and operational effectiveness of the building, individual pieces of equipment and systems (i.e., central plant, variable air volume air handlers and boxes, etc.) to understand what causes the interval data profile. The final element of this analysis is to determine what systems are in place that may be utilized. This technology provides a foundation for operational programs, and first cost for participation may be lowered by leveraging existing technology and systems in the building:

- BAS manufacturer, type, and vintage

- Centralized control of multiple locations; internet accessibility

- Communication type (Ethernet, proprietary, phone line, etc.)

- Electric metering through the BAS or dedicated energy information system

- Lighting control or automatic dimming through the BAS or standalone

The forensic energy analysis combined with the baseline and interval data history for electricity, and other energy sources, is the foundation for implementing an effective analytics program. This program can then be tuned to drive high performance energy management programs. As stated, this book is focused on the energy applications of analytics, but more importantly on electricity. This is because electricity is a highly transient energy source, which can experience significant price volatility, but also because it is the most essential fuel for every aspect of business and quality of life in the modern world. Some may ask what is meant by price volatility, because the prices have traditionally been regulated. This will be touched on in more detail throughout the book, but the simple fact is that electricity is one of the largest controllable costs for any building, and effective operations are the logical result of effectively applied analytics. Legislative and regulatory policy that defines the rules by which the electricity business operates will be touched on elsewhere in this book; the primary goal in this chapter is to discuss how analytics can be used as a tool to drive high performance building operations.

One last comment on this topic regards energy procurement, which was discussed as part of $4^{Plex}$ *Energy* but won't be covered more in this book. There are many sources for content on this topic including Association of Energy Engineers conferences and seminars. The emphasis here will be on using energy and analytics to effectively and efficiently manage the building, rather than on procurement. However, it is important to note that these tools will arm the manager with very effective information to use in negotiating the best energy procurement agreements. Particularly if the manager is able to use interval data to drive a better understanding of how their consumption and demand patterns have traditionally manifested, this will be very helpful in creating a solicitation and defining needs. Beyond that it should also be possible to negotiate with retail energy providers on terms that make the manager a better electricity customer. For example, the manager could demonstrate through the history that the building can operate within certain thresholds. A well operated building with effective analytics can operate under certain consumption/demand thresholds, even during peak period, and this will allow managers to provide the seller with a high degree of confidence on power to be provided. This is the basis for many of the operational programs run by utilities that will now be discussed. Electric utilities must serve all of the load (electricity) use on their systems to ensure that they can keep the customers' lights on and that they don't make unnecessary investments in new infrastructure. This requires that enough generation, transmission and distribution infrastructure is in place to serve customer demand. They may need to buy capacity from other utilities who have excess, are able to provide power onto the utility's grid and are willing to sell. The option is to operate programs that help them to balance bringing peaking generation units on line and orchestrated electric demand reduction. As the same time many utilities incentivize actions the customer may take that will help them in balancing load on the systems and managing demand with "base load" generation that typically runs full-time and peaking plants that can come on line quickly to meet peak demand.

**Energy Analytics in Operations**

With energy and analytics the next critical topic has to do with ensuring that the right analysis is conducted, based upon the needs and goals of the organization. As part of this analysis, the manager should pay careful attention to internal needs, as well as to programs from utilities and others. At the end of the day the goal should always be to meet organizational goals, with the minimum level of capital and operational resources, while ideally finding programs that provide ongoing funding or "electricity income." Therefore it is worthwhile to consider the programs that leverage energy and analytics data, and that also use these tools to evaluate taking actions that can improve overall building performance. Table 5-1 provides background on energy management operations, many of which that are aligned with programs that provide funding. It should be clear how valuable the energy and analytics data are in determining operations that can be implemented in a building. The table outlines four categories of operational activities for buildings, and the first column lists factors for consideration. The factors include what motivates action, what design considerations are necessary to implement the action, what the operational processes are that must be implemented, and where the action is initiated? Demand response is the only measure shown as initiated remotely, and this will be discussed further below.

There are five types of energy management operations in Table 5-1. Efficiency and conservation applies to any type of energy resource—electricity, natural gas, etc., as well as water. The other four operational types however apply specifically to electricity, and will begin to make it clear why energy and analytics technology is so important for the future.

Table 5-1. Energy management terminology and building operations

| Factors | Efficiency and Conservation (Daily) | Peak Load Management (Daily) | Distributed Generation (DG) (Daily) | Demand Response (Dynamic Event Driven) | Electric Market Participation (Analytics Based) |
|---|---|---|---|---|---|
| Motivation | Economic Environmental protection Resource availability | TOU savings Peak demand charges Grid peak | Energy Resiliency Cost Savings Optimization Carbon reduction with Solar PV | Price (economic) Reliability Emergency supply | Manage and reduce cost |
| Design | Efficient shell, equipment, systems, and control strategies | Low power design | Many options: -Solar PV -Micro-turbines -Combined heat & power | Dynamic control capability | Evaluate electric use, DG, interval data profile and building need to identify capability |
| Operations | Integrated system operations | Demand limiting Demand shifting | Balance utility purchase with on-site generation, and also may supplement heat | Demand shedding Demand shifting Demand limiting | Integrated system operations |
| Initiation | Local | Local | Local | Remote | Local |

## ENERGY EFFICIENCY AND CONSERVATION

The definition of energy efficiency is the percentage of total energy "input" that is required to achieve useful "work." So the efficiency percentage increases as less energy is required to perform work. This could be the result of installing a high efficiency motor, or a variable frequency drive on an air handler, which achieve the same work; moving air, with less energy (in this case electricity) input. Efficiency focuses on insulating the work from the energy reduction, so that the process is not impacted by the diminished energy input. Energy conservation, on the other hand, reduces unnecessary energy use. Among the first steps that should be taken in most buildings is to look for conservation opportunities, such as reducing hours of equipment operation. If equipment is being operated when the building is unoccupied, i.e. lighting, then putting programs or technology in place to ensure it is turned off, is energy conservation.

Both energy efficiency and conservation provide utility bill savings, and benefit the environment by reducing emissions from power plants. Energy efficiency measures can permanently reduce peak demand by reducing overall consumption. As noted, in buildings this is typically done by installing energy efficient equipment and/or operating buildings efficiently. Energy-efficient operations, a key objective of new building commissioning and retro-commissioning (for existing buildings), require that building systems operate in an integrated manner. Utility rebate programs are designed to drive investments in energy efficiency, such as those mentioned above, because they lower overall consumption of energy units. This allows the utility to better manage load in the short term, but it also allows them to hedge against how quickly they need to build new electricity infrastructure.

## PEAK LOAD MANAGEMENT

Daily peak load management has been conducted in many buildings for decades. These strategies are designed to minimize the impact of peak demand charges and time-of-use rates. The most common types of peak load management methods include demand limiting and demand shifting. Interestingly these strategies have been forgotten, or neglected, by many building managers in recent years, but there has been a resurgence and the value they offer is significant. Understanding interval data and building operations is critical to making the best use of these strategies, while minimizing any negative impact to occupants.

**Demand limiting** refers to shedding loads when pre-determined peak demand limits are about to be exceeded. The manager must put equipment in place, usually with a building automation system (BAS), that monitors peak demand during the time interval dictated by the rate structure and has control capability. Demand limits can be placed on equipment (such as a chiller or fan), systems (such as cooling), or a whole building. Peak demand thresholds will be programmed into the BAS, and strategies are developed and programmed to execute an action when the peak demand limit is to be exceeded. Such actions could include turning equipment off, resetting a setpoint that causes equipment to modulate to lower operational levels and draw less electricity, etc. The system will also be programmed to restore equipment to normal operation when demand is sufficiently reduced or when the peak demand interval or period has ended. Customers implement demand limiting to save money, by flattening the load shape (kW consumption profile) when the monthly a peak demand is about to be exceeded.

**Demand shifting** is achieved by changing the time that electricity is used. This is more challenging for buildings because there is very little non-essential equipment operating in most facilities. It is possible though, for example thermal energy storage is a demand shifting strategy. Thermal storage can be achieved with active systems such as chilled water or ice storage, or with passive systems such as pre-cooling of building mass. In California, **time dependent valuation (TDV)** * is also in use for building energy code compliance calculations required by the state building energy code (Title 24) to take into account the time that electricity is used during the year. Time dependent valuation (TDV) is an energy cost analysis methodology that accounts for variations in cost related to time of day, seasons, geography, and fuel type. In California, under TDV the value of electricity differs depending on time-of-use (hourly, daily, seasonal) and the value of natural gas differs depending on season. TDV is based on the cost for utilities to provide the energy at different times. TDV acknowledges that some efficiency measures reduce summer peak electric demand more than others.

---

*Brook, Martha (2011) Time Dependent Valuation of Energy for Developing Building Efficiency Standards. California Energy Commission. http://www.energy.ca.gov/title24/2013standards/prerulemaking/documents/general_cec_documents/Title24_2013_TDV_Methodology_Report_23Feb2011.pdf

## DISTRIBUTED GENERATION

Distributed generation (DG) presents huge potential for building owners. Investing in DG is motivated by many different factors. Some DG systems deploy micro-turbines and small-scale generation to reduce electricity cost for consumption and demand. Larger DG systems provide "combined heat and power" (CHP) to generate electricity and use the waste heat from that process to offset thermal needs in a building or on a campus. There has also been explosive growth in solar photovoltaic (PV) installations in recent years, driven by a desire for DG and by sustainability motivations to reduce carbon and improve the environment. In recent years, since Super Storm Sandy hit the coast of the northeastern United States, a new motivation is "energy (electricity) resiliency." Resiliency is about insulating facilities from the impacts of long-term power outages in the aftermath of storms and other events. The smart grid movement, for example, was instigated by a power outage in 2003 that cause the entire Eastern Seaboard to be without power for several days. Understanding interval data is critical to sizing DG installations and to operating them cost effectively.

## DEMAND RESPONSE

Demand response (DR) is one of the more exciting utility programs to be deployed in recent years. DR is a great example of electricity capital, or a program that provides an ongoing revenue stream to the customer, in the form of cash payments from the utility. These cash payments are typically made monthly to customers in return for agreeing to participate in the program, and to respond a preset number of times per year. DR is a dynamic and event-driven program that can be defined as short-term modifications in customer end-use electric loads in response to dynamic price and reliability information. To clarify, first "reliability information" means an event notification that is provided to the building electronically, and typically initiates automatic execution of a pre-agreed strategy. Initially DR programs allowed customers to respond manually be physically turning off equipment when an event was called, but this approach is not always reliable. Today almost all programs require automated demand response (ADR), and these programs typically involve an electronic signal to some form of automation technology, and include a BAS which automatically executes the response by turning off equipment, changing temperature setpoints, etc. These strategies are exactly like those discussed above under demand limiting and shifting, which can be utilized for demand response.

A demand response event notice typically begins with an email notification to the customer in advance (anywhere from 10 minutes ahead to day ahead) and an electronic event signal to the BAS at the start of the event which triggers the response. There are numerous types of response strategies that may be implemented, and utility programs will dictate which ones may be executed in a particular territory. Some of these strategies might be physically shutting off non-essential equipment (i.e., a fountain pump in an atrium), reducing demand drawn by equipment (i.e., change setpoint to reduce fan speed or modulate air conditioning to lower operation mode), or operating distributed generation (DG). As noted above, another option for notification is dynamic pricing. Dynamic pricing rate structures provide customers with a "discount" on the cost per unit of electricity or demand during most days of the year, except when the local utility is having difficulty meeting demand. At those times, the customer receives an electronic notification that the cost of power per unit will be increased by some multiple (i.e. 10 times normal rates). The customer then must initiate a similar form of DR strategy to protect themselves from the price impact. DR can also be accomplished with **demand shedding**, which is a temporary reduction or curtailment of peak electric demand. Ideally a demand shedding strategy would maximize the demand reduction while minimizing any loss of building services.

Demand response (DR) programs may include dynamic pricing and tariffs, price-responsive demand bidding, contractually obligated and voluntary curtailment, and direct load control or equipment cycling. Readers who are not familiar with DR programs in their utility territories may want to research them further. In closing this topic, it is important to clarify that the customer must be intimately involved in the design of these programs. The reference to "remote initiation" in the table may be alarming, but that refers to the event notification discussed above. Before an event is ever initiated there is a detailed analysis conducted, during which the customer works with the utility, and DR is provided to develop the strategies that will be executed. The owner is fully aware of what will happen after the event notification, and is able to renegotiate the contract to change or reduce the amount of participation with a coincident reduction in the DR payment. The final point is to reinforce that the DR participation contract also sets a limit on the number of times an event can be called each year and the number of hours each event can last. All of these factors should be considered by any building manager before participating. However, energy and analytics technology that is implemented, as discussed in this chapter, can take all of the guesswork out of DR. If the customer thoroughly understands their consumption and interval data profiles and effectively deploys BAS and other technology, it is possible to accompany DR with a well-orchestrated set of higher level building operations to drive real value. For example, the building owner could pre-cool the building before an event or charge a thermal storage system, so that reducing air conditioning capacity does not impact comfort. The case study chapter, Smart Grid, Smart People, outlines just such a project that was highly successful.

## ELECTRICITY MARKET PARTICIPATION

One of an owner's challenges with developing strategies for electricity is that the power business is regulated at the state level. Chapter 7 provides more background on this highly complex industry. For purposes of this discussion, the reader should research programs that exist in their utility service territory to determine if there is value. In a nutshell the idea here is that the customer sells power, which they agree not to use at a particular time, back to the utility. This may sound a bit odd, but remember that electricity is a very unusual commodity because it is created and consumed almost instantaneously, and it cannot be stored. Yes electricity can be stored using batteries and other technologies, but it cannot be stored on a scale that would allow utilities to inventory power the way that you inventory any other commodity. So this means that utilities must monitor demand instantaneously, forecast whether it is increasing or decreasing and bring on generation to meet the need. Utilities operate what they call "spinning reserves," which are power plants that are idling and ready to ramp up when demand increases. Given this highly complex environment, programs have been developed that make it possible for a customer to get paid to "not use" electricity. This reduces demand and makes it easier and less costly for utilities to keep the lights on at critical times. Again the reader is encouraged to research whether such programs exist in their utility territory and whether their properties might be candidates for participation.

The most important point for energy and analytics is still "one cannot manage what one cannot measure." Developing very detailed baseline and interval data for the building, and understanding what equipment and activities cause the consumption patterns is the price of entry for 21st century energy management. Armed with these data, the manager can evaluate every strategy that exists in the local market and develop effective optimization strategies to reduce energy use cost. Taking such action can also make the building resilient to power interruptions that negatively impact the mission, or work being done, in the facility.

# Chapter 6

# Analytics for Operations and Equipment Maintenance in Buildings and on Campuses

*James M. Lee, CEO, Cimetrics, Inc.*

## IMPORTANCE OF THIS CONTENT

Thus far in this text, the stage has been set for a changing energy market, and its implications on consumption and resource management in buildings and on campuses. For purposes of this text, the term campus is used to depict an ecosystem of buildings, which could be part of a traditional corporate or higher education campus, but could extend to community energy systems for neighborhoods, districts or entire cities. Analytics bring value in all these cases by providing the opportunity to manage and operate these buildings effectively, while maximizing the value transaction associated with consuming energy to fulfill the mission being carried out in that space. This value also impacts on the larger community, as well as the planet's environment. To accomplish these goals, equipment in the building must be operated in optimum fashion, thus balancing the energy consumed by that equipment with the work being done. The work performed by that equipment could be lighting, heating, ventilation, air conditioning, vertical transport or any number of other functions. In the end this balance between energy consumption and "work" is the definition of efficiency.

The goal of analytics is to collect and analyze essential pieces of data, to understand how to achieve the same or greater levels of work with less energy, or other resources, provided as inputs. The other resources include a wide range of "inputs" such as labor and materials for maintenance or, depending on the application, any number of other commodities and resources. Jim Lee provides great insight, in this chapter, to how this is done with particular emphasis on heating, ventilating and air conditioning (HVAC) equipment. HVAC represents approximately 40% of the energy use in commercial buildings, and in more complex facilities such as labs, hospitals and industrial plants it can be an even larger percentage of consumption. HVAC is also one of the most complex systems within any building and is made up of a large number of interdependent pieces of equipment. The efficiency of each piece of equipment relies on regular maintenance and good operations, and issues at this level can create even larger problems for the building or campus as a whole. Those problems include poor comfort or disruption in the work or "mission" being carried out in the building, but of equal concern they will result in an inefficient building with higher energy cost and a poor carbon footprint. Integral to energy and analytics is a thorough understanding of the delicate balance between energy, maintenance and mission, and this is Mr. Lee's forte.

## INTRODUCTION BY JAMES M. LEE

Since Cimetrics introduced automated building analytics in the year 2000, there have been many advancements in the field. Much of the discussion has focused on energy savings, but many applications of big data analytics are not specifically focused on energy. This chapter will discuss predictive maintenance in contrast to preventive maintenance, stressing the benefits of utilizing big data for predictive maintenance such as increased equipment life, improved reliability and lower labor cost. Though the specific focus here is energy, improvements in these areas will have dramatic impacts on the overall efficiency of the operation and will reduce energy use.

The goal of predictive maintenance is to save money and increase equipment reliability. Money can be saved by only making repairs or servicing equipment when necessary. The risk of equipment failure can be reduced by continuous, automated analysis of equipment performance in order to identify faults before they become critical. Whereas predictive maintenance was once limited to high-value capital assets, modern automation systems allow us to collect and store vast amounts of data, and low-cost computing power makes it possible to analyze that data.

## PREVENTIVE MAINTENANCE

### Definition from Wikipedia

Preventive maintenance (PM) has the following meanings:

1. The care and servicing by personnel for the purpose of maintaining equipment and facilities in satisfactory operating condition by providing for systematic inspection, <u>detection</u>, and correction of incipient failures either before they occur or before they develop into major defects.

2. <u>Maintenance</u>, including tests, measurements, adjustments, and parts replacement, performed specifically to prevent faults from occurring.

The primary goal of maintenance is to avoid or mitigate the consequences of failure of equipment. This may be by preventing the failure before it actually occurs which planned maintenance and condition based maintenance help to achieve. It is designed to preserve and restore equipment reliability by replacing worn components before they actually fail. Preventive maintenance activities include partial or complete overhauls at specified periods, oil changes, lubrication and so on. In addition, workers can record equipment deterioration so they know to replace or repair worn parts before they cause system failure. The ideal preventive maintenance program would prevent all equipment failure before it occurs.

Here are some examples of routine scheduled maintenance of equipment:

- Oil changes
- Belt changes
- Filter changes
- Linkage adjustments
- Valve seats replacement
- Steam traps replacement
- Boiler re-tubing
- Evaporator bundle cleaning
- Cooling tower water treatment chemical replenishment

Preventive maintenance has been the backbone of mechanical and industrial equipment operation for decades. When systems are constructed, the designers take note of component lifetimes, operating hours, wear parts, rated cycles and lubrication etc. Utilizing this information, manufacturers develop recommendations for maintenance to ensure effective operation and to optimize energy consumption by the equipment. Yet historically, elapsed time (run hours) has been used as the key driver for when maintenance activities should be performed. Although preventive maintenance is believed to be effective, in practice there are many shortcomings:

- The maintenance action is performed whether it is needed or not

- Labor is inefficiently deployed

- There is typically no verification of the repair action by observing system performance

- Deviation from normal operation is often limited to visual inspection

Preventive maintenance programs have frequently been automated by traditional computer maintenance management software (CMMS) packages, which require a user to understand the piece of equipment in order to create regular schedules for performing maintenance tasks. These are also coupled with spare parts inventory information, repair ticket tracking and enterprise accounting functionality. Most CMMS packages don't provide analytics capability and hence don't allow the user to gain institutional knowledge about the performance of assets over time. Furthermore, today's CMMS packages are unable to police the reliability of the repair, which is based on the skill level/training of the person performing the corrective action. In today's world of outsourced operations and repairs, how can one be certain that the repair technician knows what he is doing? By following up to analyze the data with an understanding of the models and operations of the machine or system, we can physically measure whether or not the action has been performed, and sometimes how well it has been performed.

PREDICTIVE MAINTENANCE

**Definition from Wikipedia**
    **Predictive maintenance (PdM)** techniques are designed to help determine the condition of in-service equipment in order to predict when maintenance should be performed. This approach promises cost savings over routine or time-based <u>preventive maintenance</u>, because tasks are performed only when warranted.

The main promise of predictive maintenance is to allow convenient scheduling of corrective maintenance, and to prevent unexpected equipment failures. The key is "the right information at the right time." By knowing which equipment needs maintenance, maintenance work can be better planned (spare parts, people, etc.) and what would have been "unplanned stops" are transformed to shorter and fewer "planned stops," thus increasing plant availability. Other potential advantages include increased equipment lifetime, increased plant safety, fewer accidents with negative impact on environment, and optimized spare parts handling.

PdM also benefits equipment that operates with lower life cycle energy consumption, and that can positively impact the mission being carried out within the space this equipment serves. For example, countless studies have proven that there are improvements in productivity and learning in buildings where the environment (temperature, humidity, etc.) is optimum.

A trivial example to compare preventive versus predictive maintenance would be in the area of air filters. Preventive maintenance would attempt to calculate an average life of a filter and, perhaps enabled by a CMMS, deploy maintenance staff to replace the filter at intervals that are shorter than this average life. This implies that some filters will be replaced prematurely and some filters will be replaced too late. With predictive maintenance, technology would be measuring the differential pressure across a filter. It is possible to see it load up with dirt over time (differential pressure increasing) and hence trigger a maintenance action at the right time. When the filter is replaced, the technology will record the differential pressure drop and hence verify that the replacement was done correctly. By analyzing the pressure drop across the filters over time, it is possible to better establish when and how often to change the filter and perhaps even glean information on which filter manufacturer sells a better product. Consider the benefits of this predictive approach, when the filter in question is serving a clean room manufacturing process area and any mistakes can translate to particulate contamination, production disruptions and potential product loss.

Although it is significantly more exotic than preventive maintenance, predictive maintenance is not a new topic. Historically, thermography and oil analysis have been done, temperature and pressures have been monitored, and occasionally, vibration analysis has been applied to rotating equipment. What is new is the ability to gather and process much more physical data than in the past. By using modern analytics, or "big data" approaches, which apply algorithms to system models, the effectiveness of predictive maintenance is greatly increased. Historically, predictive maintenance has been limited to individual pieces of equipment or "islands of automation," but now with big data analytics, a systems level of predictive maintenance is possible. By having big data sets from sensors all around the process and equipment, technology can build a composite view of systems operation and even correlate maintenance data to the comfort of the building or the integrity of a manufacturing process.

In the past, predictive maintenance was limited to high-value assets. What's different today is that highly skilled professionals can deploy analytics technology to automatically collect and analyze enough data so that predictive maintenance can even be applied to small end point devices. Some of these devices might include variable air volume boxes and process utility connection points, such as water for injection and compressed air connection points. The computer does the work, so automatic fault detection and diagnostics can be scaled down to the low-cost ubiquitous devices and sensors in a system.

Driven by automatic fault detection and diagnostics, these solutions can detect even minor anomalies and failure patterns to determine the assets and operational processes that are at the greatest risk of failure. This early identification of issues helps facility managers deploy limited maintenance resources more cost-effectively, maximize equipment uptime, and enhance quality.

Predictive maintenance can include:

• Observation of equipment performance drift
• Vibration analysis
• Critical process parameter monitoring
• Verification of repair (action)—by observing the data

WHAT IS REQUIRED TO DEPLOY
A PREDICTIVE MAINTENANCE PROGRAM?

There are several elements needed to deploy a predictive maintenance program. First, we need a big data, analytics, collection and analysis (condition monitoring) platform such as Cimetrics' Analytika solution, which can collect, model and perform automatic fault detection with root cause analysis. The analytics platform must comprise domain expertise so that the algorithms have a premeditated application to the system in question. The next critical element is data sufficiency—the availability of data from enough sensors, actuators and control parameters (e.g. setpoints) so that meaningful analysis can be performed. Then, equipment design information such as performance curves, rated cycles, design temperatures, and design flow rates are essential to understanding how the machine works. A system is configured by mapping the sensor data to the model and by entering static data or metadata, which describe the physical characteristics of the system. Furthermore, sensor data and equipment specifications are needed so that a model can be built defining the system we would like to maintain. This process is at the nexus of technology and technical know-how, and requires highly skilled experts who understand the equipment and the business application. Armed with the technical and business understanding, it is possible to build the system model. Unlike system simulation, this emulation of the process/equipment will take the model that is created and drive the real time data we are gathering from the real sensors and actuators through that model. After analyzing the model and real-time data with a series of algorithms, we can facilitate both equipment optimization and predictive maintenance notifications (alerts). Consider Figure 6-1 as an example of such high-level architecture, in which the Analytika predictive analytics solution is deployed.

Many equipment manufacturers have historically kept information about their design and operating characteristics proprietary. Now there is an opportunity for original equipment manufacturers (OEMs) to differentiate their products by proving a complete operating model, as well as sufficient sensors and actuators, can provide the data for predictive maintenance analytics. There has been a trend among OEMs not to include sensors on systems, because it is believed that their customers don't perceive the value. Predictive maintenance might provide the impetus to provide a new level of data sufficiency.

OEMs and consulting engineers suffer from a limited world because they do not actually operate buildings. However their system and building designs dictate to building operators what they believe adds value for long-term building operations. Given very limited direct maintenance experience among manufacturers and engineers, it is unfortunate that their roles in new building design enable them to eliminate sensors that could provide huge operational or reliability benefits. Very often these decisions are driven by "first cost" and value

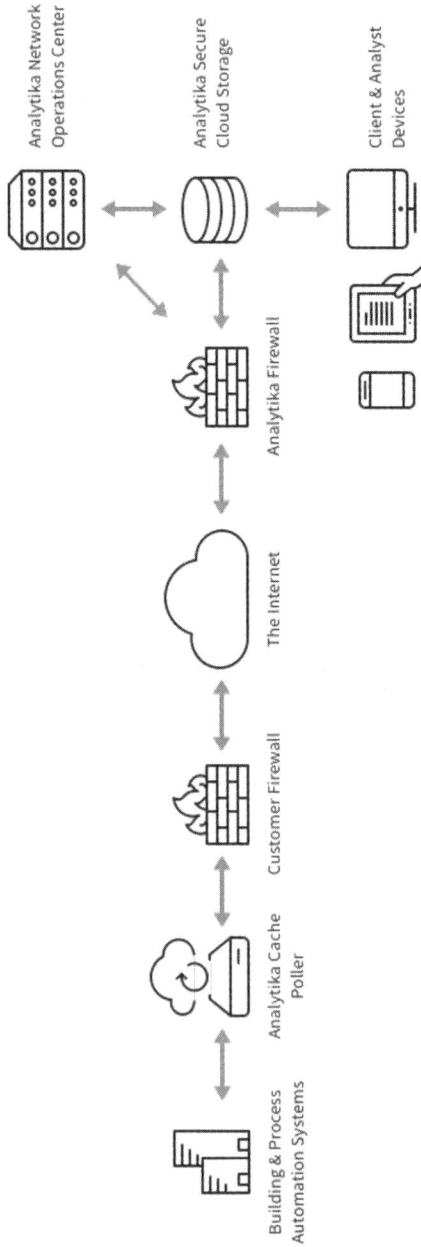

**Figure 6-1**

engineering efforts that fail to appreciate the importance of these elements of the system. A few years ago this author had a conversation with the manager at a chiller manufacturer, who said, "My guys know everything there is to know about chillers. Ask them anything about the machine, and they will tell you how to get 42-degree water. But they are completely ignorant of application- and systems-based issues, and of how their piece of equipment functions in a larger system." The purpose of this book is to educate the buildings industry, particularly building owners and managers, to the need for more active engagement in creating specifications for new buildings and equipment. This active engagement will ensure that overall access to energy and analytics technology can be simplified at the time of construction or in the future.

There are several ways predictive maintenance platforms can work. In their most remedial form, these systems can be used to collect data when the machine or system is working in steady state (regular operation) to create a baseline. Then statistical comparisons of current operation are made with the baseline to determine whether equipment performance is drifting. This is useful for detecting a fault in the system, but generally not as good at determining the root cause of the fault. With model-based automatic fault detection and diagnostics (FD&D), analytics systems maintain a preconceived model for the machines' ideal operating characteristics, and using real time data and sophisticated algorithms, analytics can predict where, and sometimes when, failures might occur (Figure 6-2).

Figure 6-3 is an example of a predictive maintenance finding from Analytika.

---

Examples of simple statistical process control techniques for predictive maintenance:
- Comparison of chiller performance curves to OEM predicitons (design curves)
- Monitoring of machine temperatures, oil viscosity, oil pressure, water pressure
- Differential pressure measurements on filters and pumps
- Pump head pressure measurements
- Vibration analysis
- Infrared thermography

---

**Figure 6-2**

| Date Opened | Priority | Status | Issue # | Category | Description |
|---|---|---|---|---|---|
| 5/2013 | High | Updated | ABC-010 | Predictive Maintenance | Equipment Rate of Travel<br><br>*Tune control loop(s) to prolong equipment life and reduce the possibility of premature equipment failure.* |

**Figure 6-3**

Table 6-1 lists the top 10 pieces of equipment with the highest actuator rate of travel during the current monitoring period. A high rate of travel may lead to premature failure of the equipment and/or the control actuator. This application is also used to identify poorly tuned control loops. Rate of travel is defined as follows:

Rate of travel:       The absolute change of a signal on an hourly basis

Relative average
    rate of travel:       The average of the rate of travel (per hour) over the monitoring period

Cycles during
    monitoring period:       The total number of actuator cycles during the monitoring period

Cycles to date:       The total number of actuator cycles since the start of monitoring

An example of a piece of equipment with a high rate of travel is shown in Figure 6-4.


BENEFITS

There are many benefits to a predictive maintenance program. Improved reliability and decreased risk of product loss and process disruptions are the most important benefits for mission critical applications such as pharmaceutical production, healthcare or manufac-

Table 6-1

| EQUIPMENT | Relative Average Rate of Travel | Cycles During Monitoring Period | Cycles to Date |
|---|---|---|---|
| VAV-157 Reheat Valve | 259 | 1,864 | 21,434 |
| VAV-150 Reheat Valve | 146 | 1,050 | 11,344 |
| VAV-223A Reheat Valve | 132 | 952 | 9,424 |
| VAV-197 Damper | 99 | 713 | 6,207 |
| VAV-North Office #1 Damper | 81 | 581 | 6,212 |
| AHU-15 Reheat Valve | 78 | 562 | 6,750 |
| VAV-North Office #2 Damper | 67 | 481 | 4,236 |
| VAV-023 Damper | 59 | 424 | 3,436 |
| AHU-02 Steam Humidifier Valve | 58 | 415 | 4,025 |
| Exhaust Fan EF-12 Damper | 48 | 342 | 3,524 |

turing. The data captured in the process of predictive maintenance analysis can be used for measurement and verification as well as providing data-historian capability for compliance reporting. Increased equipment life and an increase in mean time between failures (MTBF) can also be expected. In addition, labor savings are obtained by servicing the equipment only when necessary and dispatching repair crews with the necessary parts, thereby reducing truck rolls. Finally, in today's world, many maintenance functions are frequently outsourced to third parties. Predictive maintenance analytics allow the verification of repairs using actual operating data, allowing verification of vendor and product performance.

A predictive maintenance program aims to identify the presence of a defect in such a way as to give sufficient time for the maintenance

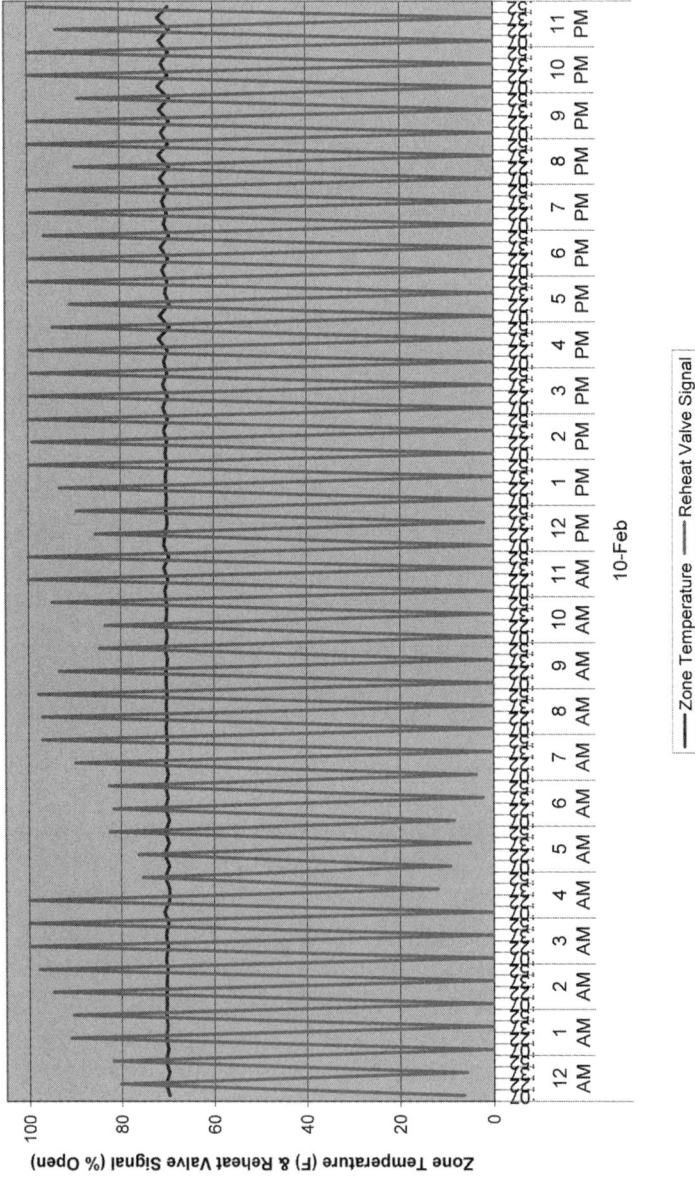

Figure 6-4

department to identify the root cause of the problem, efficiently order the parts, and schedule and complete the repair before a failure occurs. Consider Figure 6-4 which zone temperature and reheat value signal to determine if devices are operating properly.

Predictive maintenance program benefits:
- Police outsourced services
- Lower labor costs
- Increased reliability
- Risk mitigation
- Increased equipment life

**Figure 6-5**

CONCLUSION

The big data analytics revolution is beginning to enable true predictive maintenance on a large scale. Users of predictive maintenance analytics can now enjoy the benefits of cost savings, increased reliability, increased equipment life and reduced risk of process disruptions. These benefits also drive reduced energy consumption and lower energy cost. A future challenge will be the integration of predictive maintenance systems with enterprise asset management systems. Although this would seem simple at first, there is a need for a human decision maker to add business insight as to what maintenance investments should be made.

Beyond predictive maintenance, there are many other benefits to big data analytics of building systems. For example, by combining automation data with shop floor data and quality data, we can begin to gain insights that enable enterprise level risk management and process optimization. These topics and others are beyond the scope of this chapter. They highlight that we are just scratching the surface of what is possible with physical world data analytics.

Of equal importance to this book's focus on energy and analytics, is a whole new category of benefits. Any energy auditor can tell the reader that these analyses begin with a thorough evaluation of utility bills and energy consumption history. In that process it very quickly becomes evident that the explanation of consumption patterns leads to understanding how buildings and equipment are operated. During one energy audit the author conducted years ago, a fellow consultant became frustrated at how difficult it was to make the point that poor building maintenance leads to higher energy consumption and cost, and offered this analogy. The consultant expressed to the building manager that it was unlikely that anyone would maintain a $100,000 sports car as poorly as the $100,000 building chiller was being maintained, and asked why?

The power of analytics technology presents an opportunity for thorough and robust analysis of equipment and building operations to produce significant value. At the same time however, there is a diverse and complex set of functions to be carried out by these systems. This means that the reader must to be fluent in all of these functions to effectively evaluate and use analytics technology. It remains to be seen how analytics technology will be configured. Some of the initial systems on the market are one-size-fits-all, providing energy analysis, FD&D and attempting to embrace even broader feature-sets, while others offer more focused functionality. The best approach for prospective buyers is to complete a thorough needs assessment to determine both requirements and goals. With that information it will be easier to target potential solutions.

# Chapter 7

# Analytics for 21st Century Electricity Marketing and Energy Management

Energy and analytics are the next wave of technology evolution across industry. The term "big data" has received a huge amount of press, but these vast amounts of information are more the problem statement. Making sense of that data by summarizing and analyzing it to look for anomalies that point to problems and/or improvement opportunities, requires analytics and begins to unlock real value. Even more exciting, leveraging energy and analytics to unlock new market opportunities enhances value exponentially. That is really what 21st century energy management is all about. To be clear, the focus in this chapter will be on 21st century electricity management. As stated in earlier chapters, electricity is the highest quality fuel on the planet, while at the same time being the most capital intensive and environmentally impactful source of energy as well. Figure 7-1 is from the Galvin Electricity Initiative (www.galvinpower.org) and sheds more light on this fact. The "supply side" of the electricity system includes power production, and has inherent inefficiencies. That said, the electricity system is a marvel of technology. In 2003 the National Academy of Engineering in the United States published "A Century of Innovation: Twenty Engineering Achievements that Transformed our Lives." In that report the authors outlined what they believed to be the top 20 engineering achievements of the 20th century, and electrification was at the top of the list.

This book has already covered a good deal of content relating that the time is now for energy and analytics, particularly with electricity. The purpose of this chapter is to talk about how these analytics solutions may be applied to the new electricity marketplace. It will also

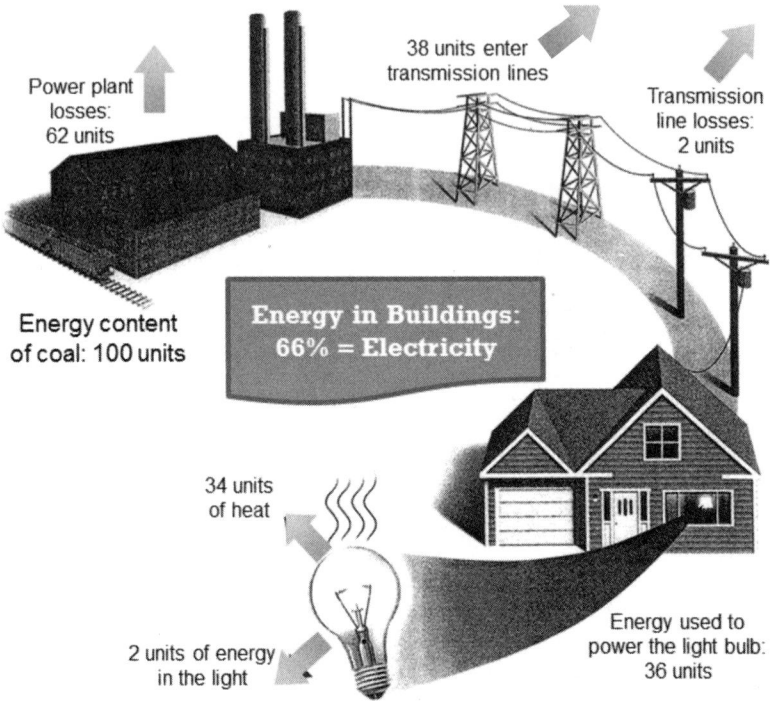

Power plant
losses:
62 units

38 units enter
transmission lines

Transmission
line losses:
2 units

Energy content
of coal: 100 units

Energy in Buildings:
66% = Electricity

34 units
of heat

2 units of energy
in the light

Energy used to
power the light bulb:
36 units

**Figure 7-1**

provide concrete examples of how this technology may be applied, and point out a few very strategic opportunities that could transform energy management in the 21st century. This dialog will begin with what the author calls the electricity triple threat. This triple threat came into focus with the June 2, 2014, Environment Protection Agency's ruling on coal-fired power plants, which was discussed in Chapter 5. The ruling is intended to tackle the carbon impact on climate change that is produced by electricity from the "plant to the plug."

This bold policy initiative may be controversial, but it puts a spotlight on an "electricity triple threat." EPA Administrator Gina McCarthy outlined the "Clean Power Plan" and its goal to cut carbon pollution from existing coal-fired power plants. The new rule mandates a 30 percent reduction in carbon dioxide emissions from these power plants by 2030. Each state was given customized goals, based upon local conditions, and the Obama Administration gave states until 2017 to develop local plans to meet those limits. In the original announcement,

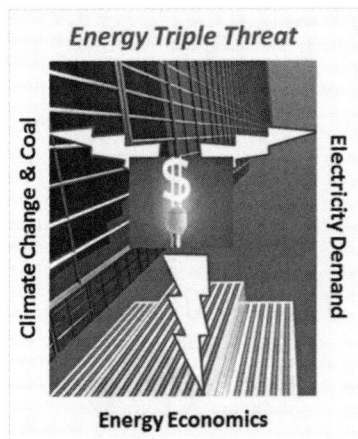

**Energy Triple Threat**

Climate Change & Coal

Electricity Demand

$

**Energy Economics**

McCarthy highlighted local success stories in cities like Salt Lake City and St. Paul, Minnesota, which the agency hopes to see replicated across the country.

So how does this plan relate to an *"electricity triple threat?"* The triple threat for electricity emphasizes that it is a very high quality form of energy, but also that is a commodity with some very unique characteristics, including that it cannot be stored in bulk. The triple threat consists of three conditions that could blindside building and energy managers:

1. **Climate Change**—The U.S. Environmental Protection Agency (EPA) initiated rules to reduce up to 30% of the carbon emissions from the 600 U.S. coal-fired power plants. These rules may force closing plants that are too expensive to retrofit, but it is also exciting because it will very likely spur investment in energy efficiency and renewable energy. In New York, for example, Indian Point Nuclear Generating Station is scheduled to be shut down in 2016. This plant provides around 20% of the electricity consumed in New York City, but the state's primary plan to replace that energy is not to build another plant. Instead efficiency is a key strategy being used by New York's Con Edison to help offset this supply-side gap. In fact, Con Edison launched a $200+ million per year program to drive that strategy in mid-2014. California utilities grappling with the shutdown of San Onofre Nuclear Generating Station are engaged in a similar strategy. The EPA of course is targeting coal-fired power plants, which could add to the challenges faced in New York, California and elsewhere. McCarthy pointed out several times in her 2014 press conference announcing the new ruling, that cities and the state have already seen proven results taking the same types of action. Regardless of the readers' beliefs about climate change, this action will dramatically affect all energy users.

2. **Electricity Demand**—With the U.S. economic recovery over the years between 2009 and 2015, U.S. electric demand saw gradual

and increasing growth. Electric utilities dodged the "load growth" bullet in the early years after the downturn due to business decline, but that changed just as the major source of electric generation (40% U.S. electricity comes from coal) was proposed to see a forcible reduction under the EPA ruling. Increasing demand for electricity can stress the grid, particularly at peak times, and that combined with potential reduction in supply sources could have an affect on energy reliability or "resiliency."

3.  **Energy Economics**—All of the changes discussed above, driven by climate change and growing electric demand, and coupled with increasing amounts of natural gas generation, can be expected to impact electric cost too. The good news is that higher electricity cost will improve the ROI of efficiency, renewable energy and distributed generation projects, but there could be real challenges in the near term. Even so, higher electricity prices may be welcomed by those who believe that energy is underpriced, and the best way to manage consumption is to raise cost. BUT, the alternative position is that our "global economy" has seen a precarious recovery that could be threatened by major price increases to a commodity that is essential to economic vitality and growth.

Without question, effectively managing buildings and their interaction with a gradually smarter electric grid, is the most effective way for owners and managers to protect themselves from the energy triple threat. To restate the obvious, energy and analytics are the best way to drive that effective building management.

Based on this brief discussion, it may be clear that the grid is reliable, but relatively inefficient, and there are some major drivers in play that are could negatively impact the electricity system. One of the most significant challenges with electricity, as an energy source and a business, is that it is regulated independently in every state. In addition, ±20 states are deregulated for electricity which can add benefits, but also makes it more confusing. The result is a complex and intricate framework of regulations for the retail electricity business at the state level, and the wholesale electricity business at the federal level. So it is important, in the context of energy and analytics, to consider how this new technology can assist energy buyers in managing electricity consumption and cost. In the 21st century, building owners will buy electricity, but they will also become much more intelligent and automated in how they use energy. At the same time, more owners will generate power at their buildings, and sell power back to the grid when it is economically feasible. All of this leads to a new reality for electricity, and even more importantly to a great need for managers to deploy analytics to optimize energy use, cost, and sale back to the grid. In Chapter 2 on *4Plex Energy* these ideas were covered, including purchasing options from the local utility or a retail energy provider. It is common knowledge that the only way to effectively manage this level of energy complexity is to baseline and benchmark buildings, as discussed in Chapter 5. As discussed, it is important to pay particular attention to "interval electric demand data" and to the occupancy and use profiles in the building that impact all of these characteristics.

Electricity purchasing is also a great example of where interval data on energy consumption (kilowatt hours per interval period) can be valuable. Having the most detailed electricity consumption analysis is a great way to negotiate the best prices. Quite simply this means that buying electricity is much more complex than contacting the local utility to turn the power on and then paying the bill each month. In closing on this topic, those who have heard this author speak on grid topics, may have heard an outline of the "urban myth" to the first great round of electricity deregulation. The myth was that more sellers will lead to lower prices, but as outlined throughout this book, electricity is a unique commodity that cannot yet be stored at scale. The belief that more sellers will translate to better prices drove the first wave of electricity deregulation. This author believes that there will be a second wave of electricity deregulation, and it will be driven by a new awareness—more sellers do not lower electricity costs, but *smarter buyers will*. Leveraging energy and analytics as outlined in this book will make buyers smarter, enable them to drive better prices and to insulate themselves from market forces and fluctuations in electricity resilience.

Electricity has been an essential energy source for over 100 years, but in the 21st century the dynamics associated with this business will demand analytics. Owners will be able to keep the lights on without this technology, but they will be leaving money on the table—perhaps large amounts of money.

The growing awareness of how complex and essential electricity analytics and management technology has become, is among the reasons that the United States Green Building Council has launched a new excellence program called PEER. PEER (Performance Excellence in Electricity Renewal) was launched at the GREEN-BUILD conference in 2014. PEER takes great strides in raising awareness of the opportunities presented by managing buildings more proactively with regard to their interaction with the electric grid. Greenbuild is the annual event put on by the United States Green Building Council (USGBC). USGBC's impact on buildings worldwide has been astounding. The LEED (Leadership in Energy and Environmental Design) certification program has changed the way new buildings are designed and existing buildings are renewed.

PEER is so critical because "technology is blurring the lines between electricity demand in the building and the power industry, driving the need for a more integrated understanding of the essential interplay between buildings and energy generation, transmission and distribution systems. PEER seeks to enhance this interconnectedness and transform the power industry by forging cross-sector, outcome-driven collaboration, and leveraging emerging technologies, policies and consumer understanding." That quote is from the USGBC announcement at GREENBUILD. More information on PEER certification at www. usgbc.org/PEER.

USGBC and the Perfect Power Institute (PPI) launched PEER at a banquet in New Orleans during GREENBUILD 2014. They invited luminaries from the green buildings and smart grid world for the launch and announcement of the first entities to receive PEER certification. The Perfect Power Institute (PPI), which was created by the Galvin Electricity Initiative, developed the PEER standard, and it has now become part of USGBC. During the smart grid era, readers may remember that Bob Galvin, former chairman of Motorola Corporation and visionary who started the cell phone industry, became very interested in electricity. This author had the honor of working with Bob and Kurt Yeager, who co-founded the Galvin Electricity Initiative, for several years. True thought leaders, these two giants of business and industry shared a vision for transforming the electricity business. Kurt Yeager is the former CEO of the Electric Power Research Institute, and it was this author's sincere pleasure to attend the USGBC banquet with him. PEER embodies much of the vision that Bob Galvin and Kurt Yeager had for electricity.

Given the significant breadth of change underway in the electricity industry, it is important to highlight the diverse nature of how the electricity business works. In the discussion of triple threat, the ideas of electricity regulatory policy, and therefore how the business works, were touched on. The first thing for any reader to do in defining their "electric reality" is to research and develop an understanding of how electricity is regulated where they own property. A good place to start is the Energy Information Administration, which publishes a status update (www.eia.gov/electricity/policies/restructuring/restructure_ elect.html).

Next, on a state-by-state level it is helpful to go to www.dsireusa. org, and then to research at the state level through the state energy office, the Public Utility Commission and with utilities themselves. Finally the reader should research the local utility and programs that support energy and analytics. Taking time to understand the regulatory environment, as well as the programs that exist, is the big picture. This information can help to frame the strategies that should be deployed for energy and analytics in any building or complex. It is possible to point to many strategies that hold promise for any building, but the specifics will come into focus based upon the local electric reality. In the context of large opportunities however, the balance of this chapter will discuss some exciting opportunities that exist globally, and that building owners and energy managers should evaluate for individual electricity markets.

The opportunities for electricity and analytics are unparalleled, compared to any other energy type. First and foremost, all managers should explore demand response and rebate programs that may be available in local utility territories. These programs can create significant opportunities to reduce near-term operating costs and generate capital to improve buildings. Such programs include on-bill financing, which is available in many utility territories. Explore www.dsireusa. org to learn more. Electricity and analytics will be instrumental in leveraging the greatest use from these programs, using approaches such as those discussed throughout this book. Integral to this discussion is metering and interval data monitoring. First it is ideal to capture electronic data directly from utility revenue meters. Most utilities will provide the interfaces for a fee. Second it is critical to be sure that the interface is providing usage (kWh consumption), demand (kW) and interval data (kW demand or kilowatt hour per sliding window billing period). Interval data on demand is the other critical data that must be captured. As discussed in this book, it is unquestioned that you cannot manage what you cannot measure and these are the data that must be measured.

This book emphasizes that discussions of electricity must first be categorized based on whether they refer to electric supply (traditionally the utility side of the meter) and electric demand (the customer side of the meter). However, the lines are truly blurring, and there is no better example than the resurgence of "distributed generation" (DG). DG is the term that will be used here, but this technology may also be called microgrids, micro turbines, onsite generation, cogeneration, combined heat and power (CHP) or solar photovoltaic (PV) power. This is a topic that most utilities loathe and, for more than four decades, it is a practice that they have discouraged. The Public Utility Regulatory Policies Act (PURPA) opened up the opportunity in 1978, and there have been several waves of implementation. Many large electricity users, particularly university campuses have deployed CHP successfully for decades. The main distinction with CHP is that it is almost universally applied in applications where all of the energy is consumed on the customer's side of the meter.

There is an "interconnect" required with the grid, and the power from the grid is intermingled with the DG electricity, but the number of universities that use the power on campus is much larger than those that sell back to the grid. The term microgrid has been used in recent years to refer to those systems that are capable of two-way power transactions, meaning to buy or to sell back to the grid. The term transactive power has also been used. This author is chairman emeritus of the GridWise Architecture Council, and that body has published content on this topic at www.gridwiseac.org. According to Gridwise, "the term transactive energy" is used here to refer to techniques for managing the generation, consumption or flow of electric power within an electric power system through the use of economic or market based constructs while considering grid reliability constraints. The term "transactive" comes from considering that decisions are made based on a value. The transaction itself may be based on the sale of electricity, or if the regulatory environment is prohibitive, it may be a purely financial transaction. An example of an application of a transactive energy technique is the double auction market used to control responsive demand-side assets in the GridWise Olympic Peninsula Project. More on The Olympic Peninsula Project may be found in the paper "Pacific Northwest GridWise™ Testbed Demonstration Projects: Part I. Olympic Peninsula Project," PNNL-17167, October 2007, Pacific Northwest National Laboratory, Richland, WA.

Solar photovoltaic (PV) systems may be the best example of the lines blurring between the supply and demand sides of electricity. The growth in solar PV installations over the last decade has been exponential, and it is not uncommon for cities, universities and others to be generating 25% to 50% of the electricity they use from PV panels onsite. Yet PV may also be the best example of a huge gap in "intelligence." DG systems, as described above, nearly always have a high degree of control and instrumentation, and have also been hot spots for early adoption of energy and analytics technology. PV, on the other hand, is almost always on a "communication island." This means that the only instrumentation on a PV system is usually the utility meter installed to capture data for renewable energy credits (RECs). These meters may use AMI technology for remote reading, but they do not communicate with anything else. This means that the opportunity to leverage renewable energy, in the form of PV, for value beyond the kilowatt hours that come from the system at any time, is often lost. For example, PV systems obviously reduce peak demand, but this author has never seen a system that actively managed the PV system and also included a demand (limiting) management strategy. Implementing energy and analytics to leverage the combined power of PV and demand limiting offers tremendous energy and economic value.

Most PV companies don't understand peak demand, or how to control it, so they exclude any evaluation of this opportunity from their analysis. They also exclude strategies to manage PV and demand from implementation plans. This is a huge mistake. Consider how much value could be gained if building automation system (BAS) technology were combined with energy and analytics technology to control the overall system. Such technology would balance three criteria: 1) optimal efficiency, 2) maximum value form PV generation, and 3) active control of building electric use, through demand limiting, when PV output drops. This combination of strategies would maximize cost effectiveness. This is the next frontier for buildings that deploy any form of DG, but particularly solar PV.

In summary, the electricity business is highly complex, and yet electricity itself remains the essential ingredient for every aspect of the 21st century lifestyle as well as every aspect of business and the economy. Energy and analytics for management of electric supply, demand and cost are untapped, and can be expected to see huge growth in the near future.

# Part II—System Technology

*Ken Sinclair AutomatedBuildings.com*

I am very pleased to provide perspective on system technology, Chapters 8 through 18, of this amazing book. I have known Jack McGowan for over 30 years. When I first met him he was writing a book, and I was honored when he used some of my material for one of the chapters. He has spent much of his life as an educator of our industry so we are all pleased that he has taken on this task of sharing his insight on the big picture of our industry, and the important topic of energy and analytics is this book. Jack has the ability to capture and organize all that is around him and contextualize it into value for others.

Jack has been a personal mentor plus the greatest fan ever of mine and Jane's efforts at creating and maintaining what is AutomatedBuildings.com.

He is the rare combination of a teacher and a doer, hence his ability to prepare a book of this magnitude, volume and embodied wisdom.

It is important to understand this book is written for people—people like you. People are our only asset, technology may come and go, but at the core of the industry are the same people who have been there for years. The problem is these core people are growing older, and much of the discussion now is that we all need to plant new people, nourish them, and help them grow.

This book does an amazing job of providing a base, the actual ground where these new younger people can grow.

Your company's and the industry's technologies may come and go, but the people are the only true assets that remain and recreate new businesses to keep the industry strong. This greatly increases the importance of the induction of new blood—younger folks with IoT smarts—into our industry. If we are to build on our existing assets, the people, then we need to invest in education and transfer of the knowledge of our assets. We need to look at new talent as an investment that can greatly increase our existing assets. As you read the following chapters, stay focused on how our assets, the people, are needed to make these technologies achieve their full potential.

Buildings connected with open protocols to the powerful internet cloud and its web services are redefining the building automation industry. The result of that redefinition is that the reach and the visibility of the industry have never been greater, nor has change been so rapid. Our "clouded" future includes new virtual connections to buildings, from the communities they are part of, with both physical and social interactions. An example is digitally displayed energy/environmental dashboards to inform all of the building's impact in real-time energy use, plus the percentage generated from renewable sources. And connections to the smart grid make buildings a physical part of their supply energy infrastructure.

The ability to operate buildings efficiently via the internet cloud from anywhere allows the building automation industry to be better managed and appear greatly simplified. Web services, or software as a service (SaaS) as it is sometimes called, coupled with powerful browser presentation are changing how we appear and interact with clients.

The data cloud for our industry has become real. As we see applications and services moved "off-site," you can imagine the opportunities for managing real estate, reducing energy and providing value-added applications for buildings.

We must unhinge our minds and find new pivot points from which to build our future. We must embrace the power of the cloud while increasing our comfort level in using the solutions within.

## Cloud Opportunities

New applications and infrastructure do not reside in end users premises; instead, the end user accesses the application on demand via a web browser on any device. This means he can concentrate on using the application for its purpose, without investing in capital expenditure, while avoiding the overhead of installation, networking and maintenance.

With the emergence of open system protocols, wired and wireless, and worldwide emphasis on energy management and sustainability, the rate of adoption of new technology by building automation vendors has increased dramatically, particularly in the use of web technology and open system architecture to integrate and converge with IT networks to create new features in a more cost effective and time efficient manner.

## Online Commissioning

Connectivity of everything is a growing reality, and with each new connection come new opportunities and new perspectives. Just as low-cost powerful connectivity is changing and actually simplifying our personal lives with internet extensions (i.e. "apps") to our handheld devices, building automation is caught up in the same connectivity growth.

In today's complex buildings, even small problems can have big impacts on building performance. Lighting, heating, ventilating and air conditioning systems need continuous performance tracking to ensure optimal energy efficiency. Yet, a formal process for data gathering and analysis is not commonplace in the nation's building stock. Plus, there's often a disconnect between the energy modeling done in isolated, one-time re-commissioning or energy audit projects, and what happens in day-to-day operations.

What's needed is a systematic approach to tracking energy utilization that helps detect problems early, before they lead to tenant comfort complaints, high energy costs, or unexpected equipment failure. That's why new robust energy monitoring technologies and monitoring-based commissioning (MBCx) techniques are now at the forefront in building energy management.

The continuing question is how to convert data into meaningful information that is contextual and actionable. The operations center is an environment where meaningful information can be extracted and presented to produce a high level of situational awareness, align related work processes, minimize workload and errors, enhance task performance, and provide information and reporting tools required to manage the building's operations.

I am very pleased that Jack has included a chapter on Haystack Connect and the Next Generation of Energy Standards, and a Chapter on The Internet of Things. It shows how this far-reaching book has information created only a few years ago and provides connections (words to Google) to evolving online resources.

When you have read this book completely, you will grasp the scope and complexity of *Energy and Analytics – Big Data and Building Technology Integration*. Please share your new-found knowledge with your peers in discussions and on social media. Reach out to involve the people who are rising to be the new assets of our industry. Teach them what Jack taught you in this book, become their mentors, lead them to the resource of this "Theory of Everything Book" for our industry.

# Chapter 8

# Building Systems Technology: The Foundation of Analytics

A recurring assertion in this book is that there is significant potential for analytics. Yet there has also been emphasis on the importance of providing quality data to the analytics technology, and the fact that it is extremely costly to build entire new networks of instrumentation for this purpose. As a result it becomes very important to identify existing systems that have access to necessary data, and devise strategies to extract those data and make them available to analytics tools. Well, that is easier said than done. That is why the next several chapters in this book will provide in-depth coverage of the building systems that house important data. These systems include building automation systems (BAS), sometimes called energy management systems (EMS), and other building systems. It is the author's hope that the reader read the independent foreword that was provided for this section of the book. It was written by one of the most respected, and knowledgeable, professionals in the building technology, Ken Sinclair. Ken had a distinguished career as a consulting engineer in British Columbia, specializing in building systems, before starting the most prestigious electronic magazine in the industry, www.automatedbuildings.com. With automatedbuildings. com, Mr. Sinclair has created the "go to" resource for professionals seeking content on buildings, BAS, systems deployed throughout facilities and communications, and networking/web-based technology that is utilized by and interfaced with those systems. With Mr. Sinclair's introduction and deep technical knowledge, it is the authors' intent to expand the reader's view of energy and analytics to embrace building systems. Therefore this section of the book begins with a recognition that this may be new content for many readers, but also with the recognition that these systems are essential to implementing a robust and effective analytics program. It is of particular import that these systems will also provide implementation tools to enable recommendations that come from energy and analytics tools. It cannot be overemphasized that deployment of analytics requires an interdependence with building systems. Further this interdependence goes beyond the initial implementation, but requires ongoing diligence by management staff to ensure that the flow of quality data is uninterrupted, a topic that has been referred to as "data quality."

This series of chapters begins with an outline strategic planning and enterprise energy management from a team at the University of Louisiana in Lafayette. Though focused on energy management, this document highlights the need to inventory all building equipment including lighting, HVAC, etc., as discussed in Chapter 5. This type of rigorous inventory, energy history and planning process is essential to understanding the full breadth of the facilities involved, and to clearly understand the data points that will be needed. This will allow managers to determine whether data points may be accessed from existing systems. The section will go on to provide a detailed treatise on building automation and other facilities systems, as well as the underlying data communication systems that this technology deploys. All of this information is critical to avoid the need for implementing completely new sensor networks. The section will then delve into the topics of integration and IT/middleware technology. Accessing data from disparate systems requires specialized techniques and technologies that readers should understand well enough to evaluate proposals from solution providers.

As discussed in the author's note and in the initial chapters of this book, this is a complex topic that is impacted by wide ranging policy, business and technology trends. Energy management in the 21st century will require energy and analytics technology. Armed with this technology, building owners/managers and certified energy managers CEMs will be able to fully optimize their buildings and campuses. The goal of this series of chapters, and Section II of this book, is to assist the reader in this process of optimization, by providing detailed content regarding the fabric of underlying building and energy technology that is critical for analytics to drive maximum value. Ken Sinclair's foreword along with this chapter are here to provide insight on how these technologies are interconnected and interoperable, and what that could mean for energy and analytics. These underlying systems must exist to capture enough data for meaningful analytics. As noted, in many cases they are also required to execute the changes recommended by analytics engines.

Chapters in this section will discuss applications that are driven by the analytics but will also address new technology and practice. One such important technology evolution discussed is the Internet of Things (IoT). IoT is most appropriate because all of the technology discussed in this book has become interconnected with communication networks, and the internet. Access to the internet makes it possible for analytics to enable and automate smart decision making. This will impact energy and buildings in a profound way. It will be similar to the impact that the internet has had on consumers, enabling them to explore options, and compare and select products and services, in ways unlike at anytime in the past. Another important topic that could unleash even more profound market growth is data modeling and the exciting work of Project Haystack. Haystack represents the efforts of numerous emergent technology companies that have chosen to band together and tackle a major challenge of analytics. That challenge is devising data standards that make it possible to capture data from myriad systems quickly without the need for mapping and converting points. Finally this section will touch briefly on the topics of financing and measurement and verification for energy projects. Whether the financing discussion focuses on mechanisms that may be used to acquire systems, or on measurement and verification to ensure that these system investments perform as intended, it is an important topic for readers to consider.

# Chapter 9

# Six-Sigma Approach to Energy Management Planning

*Jim Lee, Kirkrai Yuvamitra, Theodore A. Kozman, Kelly Guiberteau*
*Department of Mechanical Engineering*
*University of Louisiana at Lafayette, Lafayette, Louisiana 70504*

## INTRODUCTION

The first step for any energy and analytics program is planning. It may seem out of order therefore to provide this content as Chapter 9 of the book. This is because the first section of the book was designed to provide introductory content that outlined the full breadth of market and industry potential that exists for energy and analytics. If given that content, readers determine that it is time to implement energy and analytics, this section will be very useful in providing an understanding of what is required. This chapter from a team at the University of Louisiana at Lafayette is highly effective, and equally exciting as it incorporates the idea of Six-Sigma. At its core the idea behind energy and analytics is to create a framework that optimizes systems performance for both efficiency and operation effectiveness. Any program of this type must be able to accomplish those goals in a fashion that is repeatable. The beauty of integrating Six-Sigma is that it underscores that continuous improvement is essential, and also provides a framework that has been highly effective in analyzing processes and developing improvements. It is particularly poignant for this author, because I had the opportunity to work with Bob Galvin, in many ways the father of Six Sigma, until his death. As discussed in Chapter 7, Bob founded the Galvin Electricity Initiative with Kurt Yeager because he believed that the power industry was in dire need of an excellence program. He believed that the regulatory business model stifled creativity and innovation and was to the detriment of all electricity consumers.

There are many benefits to implementing a Six Sigma energy management program, but perhaps one of the greatest is that it is built on a framework that drives it to continuously improve. This is the ideal embodiment of a solution to the old adage "one cannot manage what one cannot measure." Building a program of this type ensures measurement, but equally important it ensures an ongoing process that continues to measure and evaluate to ensure that operations are optimized, and to identify additional steps that can be taken to further enhance the building, campus, etc. In an ideal world, the purpose of energy and analytics is to provide tools for automating analysis and evaluating performance. The next logical step is to develop an implementation program to deploy building improvements, and there again the analytics enable managers to evaluate the performance of those improvements. At each stage analytics provides timely information to make timely and effective decisions about performance and actions that are necessary. The reader should note that this planning process must also consider more than efficiency. The planning process must consider the electricity demand side, or consumption characteristics described here, as well as the electricity supply side. The electricity supply side has historically been the purview of utility companies, but as discussed in Chapter 7 the rapid advance of distributed generation (DG) at the customers premise is changing this reality. DG is increasingly important to the planning process, and could include: solar photovoltaic panels on the roof, a micro-turbine or full-scale combined heat and power (CHP). Regardless of its form, DG introduces significant change to energy management planning. The reader should incorporate that thinking into the planning process. As a final note, the content in this paper clearly refers to a manufacturing process application, and as such does not incorporate all of the systems that might be encountered in a commercial building or university campus, particularly cooling equipment. However, the focus here is more specifically on the process rather than the application, and the reader is encouraged to develop any additional information that might be appropriate for their given application.


ABSTRACT

Most companies would like to reduce costs, and at the same time the government and society are pushing for more "green" practices.

By having an energy management plan, a company can reduce energy usage and reduce operating costs because energy costs money. The main purpose for an energy management plan is to decrease energy consumption and costs. This paper will cover Six-Sigma based energy management planning procedures, to include: define, measure, analyze, improve, and control. An overview of some of the major energy consuming equipment found in most industries is discussed and different energy saving opportunities are presented in the plan. The benefits from this research will provide information and a clear understanding for establishing an energy management plan and what is to be expected when performing an energy audit to save energy and cost.

*Keywords:* Energy management plan, Six-Sigma approach, energy savings, energy audit, energy saving technology, ISO 50001.

## INTRODUCTION

Energy management plan (EMP) is an important summary of activities that entails the information and steps that should be taken by a company to save energy and costs. The activities suggested are highly cost-effective requiring very little budget [1]. The EMP reported in this chapter was developed after an energy audit was performed by Louisiana Industrial Assessment Center for an oil service industry.

The objectives of an EMP are to improve energy efficiency, reduce cost, and conserve natural resources. The specific goals are:

- Continuously improve energy efficiency by establishing and implementing an effective energy management plan while providing a safe and comfortable work environment.

- Suggest an action plan with economic analysis to reduce energy cost.

- Encourage continuous energy conservation by employees through work and personal activities.

A Six-Sigma approach to define, measure, analyze, improve, and control is used to describe the five main steps in the energy management plan [2]. The development of EMP can be considered as the first step to achieve ISO 50001, a world-class energy management standard that integrates with ISO 9001 and ISO 14001 [3].

While an EMP is used to identify energy saving opportunities and tools, ISO 50001 is designed to help companies evaluate and prioritize the implementation of energy-efficient technology and promote efficiency throughout the supply chain. Superior Energy Performance (SEP) [4] is an ANSI/ANAB-accredited certification program that builds on ISO 50001 to provide industrial and commercial facilities with a pathway to continuously improve energy efficiency while boosting competitiveness. Demonstrations of SWEP can be found in [5].

This chapter focuses on use of the six-sigma approach in energy management planning. Figure 9-1 shows a flow chart for the Six-Sigma approach and the major steps are discussed in the following sections.

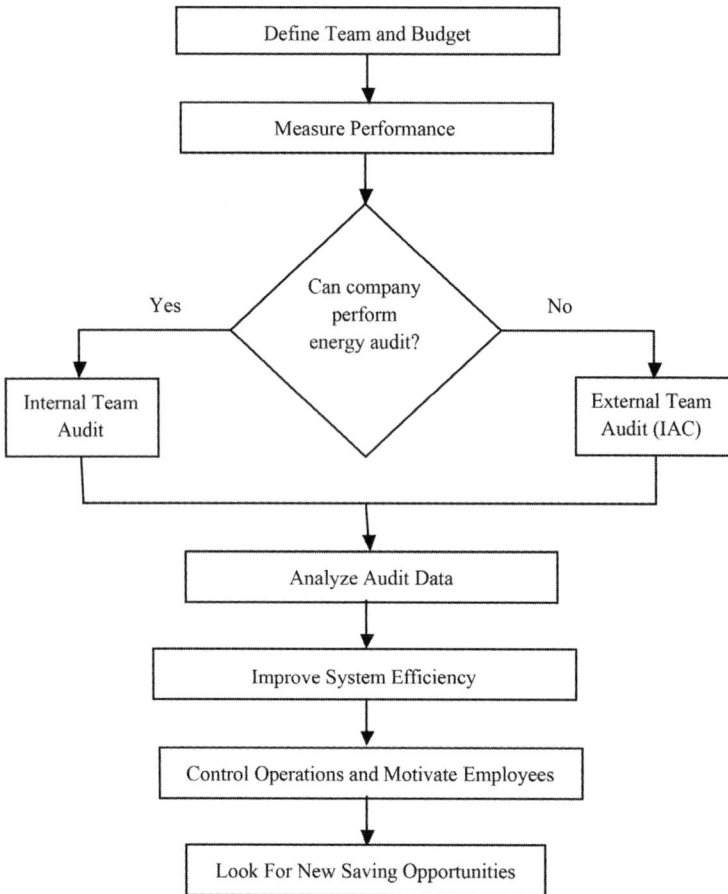

**Figure 9-1. Energy Management Plan Flow Chart**

## DEFINE TEAM AND BUDGET

The first step in the energy management plan is to establish an energy team and to budget for implementation. They are discussed below.

### Establish Energy Team

Establishing an energy team helps to integrate energy management by measuring and tracking energy performance, analyzing, planning, and implementing specific improvements. The size of the energy team will vary depending on the size of organization. In general, the energy team members are as below:

- **Energy Director**: An energy director should be a member of senior staff and have both a technical background and familiarity with the organization. Most importantly, an energy director must understand how energy management helps the organization achieve its objectives and goals. The responsibilities for energy director are:
  — Directing and coordinating the overall energy management operations.
  — Acting as the point of contact for senior management.
  — Creating the formal action plan.
  — Establishing and directing the energy team.
  — Providing budget for implementation.
  — Measuring, analyzing and monitoring progress.
  — Establishing recognition and reward program.

- **Project Engineer**: The project engineer should have a technical background and provide the team with technical specification information on production equipment. He/she is also responsible for researching new technologies, which can help improve energy savings.

- **Maintenance Supervisor**: The maintenance supervisor is responsible for providing operations scheduling information to the team to assist scheduling in the action plan.

- **Department Representative**: The department representative will cooperate with the maintenance supervisor for implementation.

### Provide Budget for Implementation

Reasonable budgeting is a must for implementing an energy management plan. This should be made available under company guidelines. Budget allocation should be known to the energy team at the start of the year. According to the data provided by the Louisiana Industrial Assessment Center [6], the average implementation budget for a mid-sized manufacturer is around $40,000 to cover insulation, maintenance, and installation of energy-saving devices. The flowchart for the define step is shown in Figure 9-2.

MEASURE PERFORMANCE

Understanding how energy is used in the past and present will help the organization identify energy performance and saving opportunities. In this step, the method for measuring performance is energy audit. Energy audit can be performed either by an internal audit team(s) or by an outside service. An outside service like those provided by Industrial Assessment Centers throughout the US will look at basic energy consump-

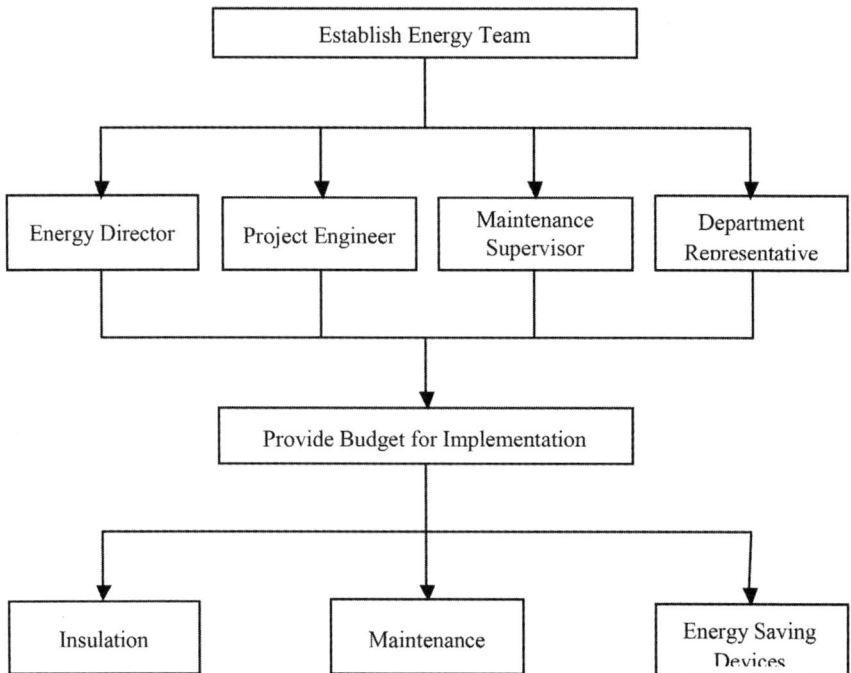

**Figure 9-2. Define Team and Budget Flow Chart**

tion practices and will not often get involved within specific processes key to the manufacture or maintenance of a product.

The advantage of an internal audit is that the energy team would be more familiar with the manufacturing process and more comfortable making suggestions for reductions of energy within a specific process. The main disadvantage of an internal team is being too comfortable with current practices, and the team will sometimes miss obvious opportunities for improvement. Another disadvantage of an internal audit is a lack of experience or knowledge of means of reducing energy.

The best way to perform an energy audit is to initially have an outside team come in and give an internal energy team the stepping stones of where to start reducing energy. There are two steps in an energy audit: pre-audit data collection and onsite technical audit. See the tables below.

**Collect Pre-energy Audit Data**

A pre-energy audit is the preparation step for a technical energy audit session, which includes collecting basic data on utilities' bills (types of energy usage, amount of energy usage, energy costs), types of equipment (size, hours of operation) involved in the processes, waste and recyclable, and production data. The data gathered during this step will allow the energy team to identify what they need to be looking for and what data they must gather while walking through the plant. A sample data collection form for pre-energy audit is shown in Table 9-1.

**Perform Technical Energy Audit**

During the technical energy audit, all real-time data are recorded and any missing information from the pre-energy audit is gathered. This step is a detailed audit of the entire facility. The data gathered during this part of the audit will be analyzed later for saving opportunities and implementation. The focus areas during the audit are:

- Heat loss for boilers

- Cooling equipment specifications and  peration time and settings for HVAC systems

- Lighting count: including type, how it is being used and lighting intensity (lux)

- Air decay test and operating pressure for air leaks in compressed air systems

### Table 9-1. Sample of Pre-energy Audit Data Collection Form

| Company: ABC Manufacturing | Plant Supervisor: John Crisler |
|---|---|

| **Energy Data** ||
|---|---|
| **Energy Usage per Year** | 19,152,000 kWh/yr (Electricity) and 4,227 MMBtu/yr (Natural gas) |
| **Energy Cost ($/unit)** | 0.05353 $/kWh (Electricity) and 11.63 $/MMBth (Natural gas) |
| **Energy Tax ($/unit)** | 0% (Electricity) and 0% (Natural gas) |
| **Energy Cost ($/yr)** | 1,025,232 $/yr (Electricity) and 49,152 $/yr (Natural gas) |
| **Comment:** | |

**Plant Production Data**

| Shift | Time | Hours | Days/Week | Weeks/Year | Hours/Year |
|---|---|---|---|---|---|
| Shift 1 | 07.00 am – 3 pm | 8 | 5 | 52 | 4080 |
| Shift 2 | 3 pm – 11 pm | 8 | 5 | 52 | 4080 |
| Shift 3 | - | - | - | - | - |
| **Comment:** | | | | | |

**Equipments**

| Types | Location | # Units | Size | Operating Condition | Time of Operation |
|---|---|---|---|---|---|
| Boilers (gas) | Boiler room | 1 | 150 Hp | Steam pressure 75 psi 250 F | 7 am – 11 pm |
| Compressed air | Outside | 1 | 185 HP | 120 psi | 7 am – 11 pm |
| HVAC | Office | 3 | 24,000 Btu | 72 F | 7 am – 4 pm |
| **Comment:** 1 HP of boiler is equal to 33,475 Btu/Hr | | | | | |

As an example of data required, a sample data collection form for boilers is shown in Table 9-1. The flow chart for the measure performance step is shown in Figure 9-3, following Table 9-2. An external technical energy audit for Cameron was performed by Louisiana Industrial Assessment Center. Future work will entail performing internal energy audits to identify new saving opportunities.

ANALYZE AUDIT DATA

Analyzing data collected from the audit can help an organization gain better understanding of the current and future energy usage and

**Table 9-2. Sample of Technical Energy Audit Data Collection Form for Boilers**

| Company: ABC Manufacturing | | | | Plant Supervisor: John Crisler | |
|---|---|---|---|---|---|

| Boilers Data | | | | | |
|---|---|---|---|---|---|
| **Type** | **Location** | **Boiler Number** | **Size** | **Time of Operation** | **Operating Condition Include Operating Temp** |
| Gas | Boiler room | 1 | 150 HP | 7 am – 11 pm | 75 psi 250 F |

| Infrared Pictures | | | | | | |
|---|---|---|---|---|---|---|
| **Picture No.** | **Boiler Number** | **Hot Spot Area (ft²)** | **Hot Spot Length/Surface Area (ft)** | **Hot Spot Location** | **Hot Spot Temp (F°)** | **Ambient Temp (F°)** |
| 1 | 1 | 25 | 2.36 | valve | 126.9 | 80 |
| 2 | 1 | 3 | 0.75 | Firebox cover | 333.8 | 80 |

**Comment:**
1 Hp is assumed to be 34.5 Lbs/Hr of saturated steam flow rate at 0 psig and 212 F°
Focus on any surface temperature that is over 140 F° (Based on OSHA Safety Standards)

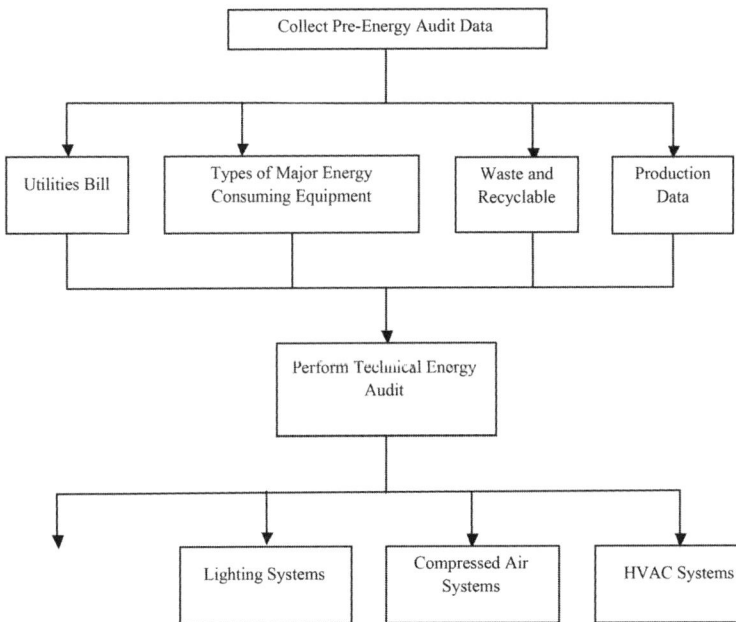

Figure 9-3. Measure Performance Flow Chart

required budget for an implementing energy management plan. Saving opportunities will be identified, and energy savings, cost savings, implementation costs and payback period will be calculated. The results will later be used for developing energy saving goals and developing an action plan. Table 9-3 summarizes the results of the analyses.

### Identify Saving Opportunities

The first step in analyzing audit data is to identify saving opportunities. Table 9-3 shows a list of saving opportunities based on an oil service industry audit. For example, based on OSHA safety standards and technical audit data, the saving opportunity for a boiler is to install boiler insulation.

A flow chart for analyze audit data is shown in Figure 9-4.

**Table 9-3. Saving Opportunities, Resources Saving Per Year, Energy and Cost Savings, Implementation Costs and Payback Period**

| # | Saving opportunities | Energy savings per year | Cost savings ($/yr) | Implementation Costs ($) | Payback period (yr) |
|---|---|---|---|---|---|
| 1 | Insulate surface for hot processes | 13 MMBtu | $145.00 | $356.00 | 2.45 |
| 2 | Preventive maintenance for group lighting replacement | 3,424 labor hours | $102,711.00 | $10,348.00 | 0.1 |
| 3 | Repair leaks in compressed air system | 324,606 kWh | $17,376.00 | $12,200.00 | 0.7 |
| 4 | Reduce number of lights | 89,463 kWh | $4,789.00 | $5,670.00 | 1.18 |
| 5 | Install Vending Mizers | 17,472 kWh | $935.00 | $1,838.00 | 1.96 |
| 6 | Insulate surface for cold processes | 645 kWh | $35.00 | $56.00 | 1.62 |
| 7 | Utilization of occupancy sensor lighting controls | 11,311 | $606.00 | $4,020.00 | 6.64 |
| 8 | Replace lighting with more efficient lights | 6,839 kWh | $366.00 | $3,220.00 | 8.8 |
|  | Total | - | $126,963 | S37,708 | - |

**Figure 9-4. Analyze Audit Data Flow Chart**

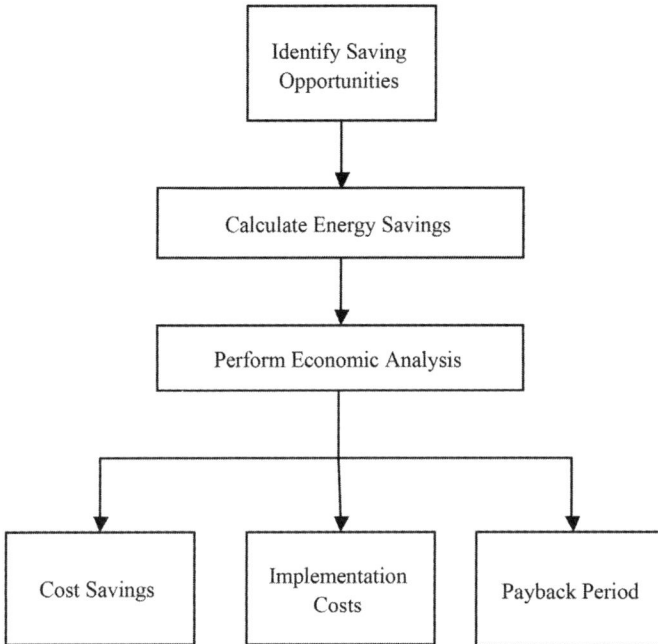

## Calculate Energy Savings

Based on the saving opportunities, the next step is to calculate energy savings. For example, installing boiler insulation can be estimated using DOE tool 3E Plus with the following input parameters:

- Indoor wind speed 5 mph
- Mean surface temperature
- The ratio of length and area of hot surface
- Insulation thickness 1 inch (fiberglass foam, 450F MF BLANKET, Type II)

## Perform Economic Analysis

The economic analysis includes estimating the cost savings, the implementation costs, and the payback period. The cost savings can be either energy cost savings or labor cost savings. For example, the cost saving for installing boiler insulation is calculated by energy savings multiplied by energy cost per unit.

Implementation cost is calculated based on the cost of material (capital cost) and the labor cost required for each saving opportunity.

For example, implementation cost for boiler insulation is calculated by total number of insulation needed and the labor cost. The total number of insulation needed is based on the total area of hot surface. The labor cost is calculated based on assumption that one labor hour can install insulation 3 ft$^2$. A simple payback period can be calculated by estimating the costs of implementation and dividing by the cost savings.

IMPROVE SYSTEM EFFICIENCY

Once the audit data have been analyzed, the next step is to develop an energy saving goal and an action plan. The improve system efficiency flow chart is shown in Figure 9-5.

**Figure 9-5. Improve System Efficiency Flow Chart**

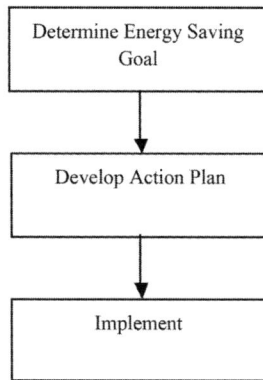

**Determine Energy Saving Goal**
A performance goal drives energy management activities and promotes continuous improvement. The goal must be clear and measurable. The goal for energy savings is normally between 15%-5% at the end of the first year. The following year's goal should be between 10%-3%.

**Develop Action Plan**
After the goal is in place, a detailed action plan must be developed for implementing recommendations. It is imperative that the energy team procures all necessary items (including budget) to achieve the implementations. The action plan must include the performance target, time line and department roles for implementations. The action plan

should start with the safety issues and then move to opportunities with the highest potential cost savings. A sample action plan may include the following activities:

- Insulate surface for hot processes (approximately 10 labor hours).
- Preventive maintenance for group lighting replacement (approximately 159 labor hours).
- Repair leaks in compressed air system (approximately 305 labor hours).
- Reduce number of lights (approximately 189 labor hours).
- Install vending mizers (approximately 1.5 labor hours).
- Insulate surface for cold processes (approximately 1 labor hour).
- Utilization of occupancy sensor lighting controls (approximately 34 labor hours).
- Replace lighting with more efficient lights (approximately 15 labor hours).

**Implement**

After the action plan has been clearly stated, the next step is to start implementing. People can make or break an energy management plan, so it is important to gain the support and cooperation of key people at different levels within the organization.


CONTROL OPERATIONS AND MOTIVATE EMPLOYEES

In energy management, there must be a control plan to monitor the progress of the work done for implementation, make the implementation sustainable, and also monitor the savings incurred. The monitoring progress can be performed by periodic monthly reports, which can show progress on any work being done, and monitoring the electrical bills each month can also show a drop in electrical usage. The control operations and motivate employees flow chart is shown in Figure 9-6.

**Perform Preventive Maintenance**

After implementation, it is important to sustain the improvement and to prevent future failures by preventive maintenance. The recommended activities for various equipment are discussed below.

**Figure 9-6. Control Operations and Motivate Employees**

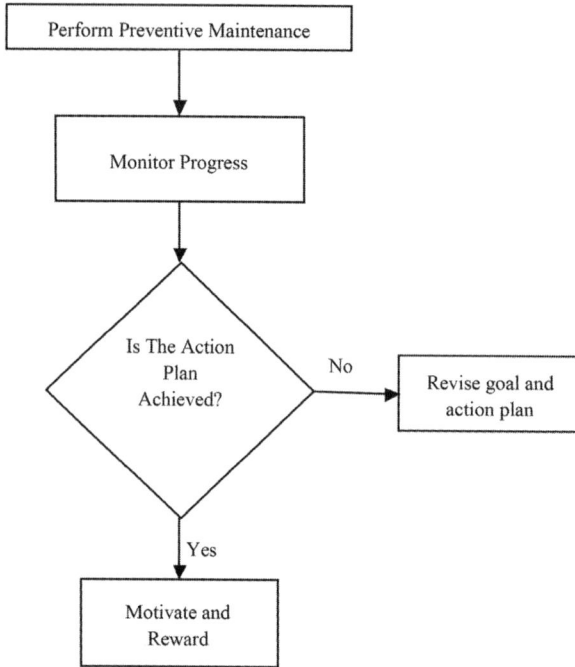

```
         ┌─────────────────────────────────┐
         │  Perform Preventive Maintenance  │
         └─────────────────────────────────┘
                         │
                         ▼
         ┌─────────────────────────────────┐
         │         Monitor Progress         │
         └─────────────────────────────────┘
                         │
                         ▼
                    ◇ Is The Action ◇
                      Plan              ── No ──▶  ┌──────────────────┐
                      Achieved?                    │  Revise goal and │
                    ◇             ◇                │   action plan    │
                         │                         └──────────────────┘
                        Yes
                         │
                         ▼
              ┌──────────────────────┐
              │    Motivate and      │
              │      Reward          │
              └──────────────────────┘
```

*Boiler Preventive Maintenance*

• Visually inspect all refractory surfaces, such as the boiler door, daily.

• Clean the surface if the fireside or waterside of boiler is dirty.

• Make sure that the air duct supply is clean, weekly.

• Check the fuel filter weekly to make sure that there will be enough fuel supplied for higher loads.

• Check wiring, monthly.

• Check weekly for leaks along the supply lines.

• Keep a daily record of fuel consumption and operating conditions.

*Compressed Air Systems Preventive Maintenance*

• Determine the amount of air leaks present in a system using a decay test. The test can be done by measuring the drop in air pressure in an air system over a period of time (about 120 seconds).

- Find and fix air leaks as they occur in the system. Air leaks can be found through a bubble test for small air leaks or though acoustic leaks. Once leaks are fixed the air pressure can be lowered.

- Operate system at the optimum pressure requirement, usually 100 psi.

- Turn off air system when compressed air is not in use.

*Lighting Systems Preventive Maintenance*

- **Cleaning of Lenses or Skylights**: Cleaning allows more light to enter the work space and reduces the need for excessive lighting fixtures. This could also be considered a safety hazard if dust collects on the lens of a light fixture and will be ignited by the heat of the fixture.

- **Group Lighting Replacement**: In most production areas high ceilings are used along with metal halide bulbs, which require the use of a man-lift to access the bulbs for changing. Group lighting is when lights are sectioned into four groups and every 6 months a maintenance team would change all of the lights in a section, regardless of whether the bulb is burnt out or not. The maintenance crew would not return to replace these bulbs until their recommended life (about 2 years) is spent. The time saved in man-hours by doing this type of light replacement is greater than the cost of throwing away some light bulbs that are not burnt out.

*HVAC Systems Preventive Maintenance*

- Inspect and refill refrigerant fluid to required level every 6 months
- Clean dirt and sludge from evaporator and condensate coils, annually.
- Lubricate fan motor and fan blade, annually.
- Inspect compressor and tubes for leakages, annually.
- Check all connectors, breakers, and wires for damage, annually.
- Clean or replace air filter, quarterly.

**Monitor Progress**

There are many ways to monitor progress of the work done for implementation as described below:

- Review weekly report for the progress of implementation.

- Review monthly report for the savings occurred.
- If the action plan is not achieved, revise the goal and action plan.
- If the action plan is achieved, go to step 5.3 and look for new saving opportunities.

**Motivate and Reward**

Offering incentives for energy management is one way to create interest in energy initiatives and foster a sense of ownership among employees. There are many ways to motivate employees such as [7]:

- **Internal Competition**: Use tracking sheets, scorecards, etc. to compare performance of similar facilities and foster sense of competition.

- **Recognition**: Highlight and reward accomplishments of individuals, departments, and facilities.

- **Financial Bonuses and Prizes**: Offer cash bonuses and other rewards if goals are met.

CONCLUSIONS

The purpose of having an energy management plan is to save energy and costs while a company still produces the same quality and quantity of products. Energy savings can come as a result of managing existing systems or from using new technologies. There are several areas in which most companies can save money such as lighting, HVAC, and production. These saving opportunities are identified through energy audits and implemented through preventive maintenance, replacement of old technology, and equipment workload balance.

As part of a company's initiative to lowering their costs, it is important to have an energy management plan in place along with an energy management team. For the EMP, the team must develop management commitment, energy audits, implementation, tracking of implementation, and a rewards system. Having a well-organized and devoted energy team is important in making progress on energy management.

**References**
[1]    Chemical Industry Digest. 2007. Effective Energy Management. http://www.energy-managertraining.com/Journal/18072007/EffectiveEnergyManagement.pdf

[2] Summers, D. 2011. "Lean Six Sigma: Process Improvement Tools and Techniques": *Prentice Hall*. 1st ed.

[3] US Department of Energy. 2012. ISO 50001 Energy Management Standard. http://www1.eere.energy.gov/energymanagement/.

[4] Achieving Superior Energy Performance. 2012. Overview. http://www.superiorenergyperformance.net/index.html.

[5] US Department of Energy. 2012. Superior Energy Performance Demonstrations. http://www1.eere.energy.gov/manufacturing/tech_deployment/sep_demonstrations.html

[6] Kozman, T. and J. Lee, 2010. "The Results of More Than 250 Industrial Assessments for Manufacturers by the Louisiana Industrial Assessment Center for the Past Ten Years," Proceedings 2009 Industrial Energy Technology Conference (IETC). ESL-IE-10/11. New Orleans, LA.

[7] Energy star, 2012. Guidelines for Energy Management Overview [Online] Available at http://www.energystar.gov/index.cfm?c=guidelines.guidelines_index

# Chapter 10

# State-of-the-art Building Automation Systems (BAS) Executing Direct Digital Control

Reviewing the breadth of building automation system technology executing direct digital control ($BA^{DDC}$) in one chapter, is daunting if even feasible. With that caveat therefore, the assumption here will be that the reader here will have, at least, passing familiarity with heating ventilating and air conditioning (HVAC) equipment, basic temperature and energy controls, and building automation systems (BAS). The topic of the book is energy and analytics, and the focus is on buildings and campuses, so it is hoped that this content will be useful to the reader. In the alternative, some readers might ask why this book covers building automation technology, anyway, and the answer is simple. As introduced in Chapter 8, among the most critical underlying technologies to successfully implementing energy and analytics in buildings, is the BAS. Data analytics topics are very popular in discussing all manner of complex infrastructure systems and are rapidly being deployed for public safety and a host of other applications. The single largest obstacle to wide-scale deployment of analytics across all building applications related to energy and operations, is the high cost of deploying smart sensor networks. Sensors are necessary to access the data necessary to perform the analytics and drive knowledge for management decision making. The important point for energy and buildings is that there are already extensive networks of sensors connected to BAS, meters and a host of other systems currently in place. Beyond that, the industry has also taken huge strides toward implementing "open systems" and migrating data to web based tools of all types, which means that this industry has already overcome the single greatest obstacles faced by so many others.

With that introduction, the reader should understand that the BAS is an enabling technology for energy and analytics in buildings. Therefore a brief historical perspective on BAS systems is warranted, followed by an overview of the technology. It is essential to discuss direct digital control which is executed by the BAS and to touch on critical market developments while outlining general system technology. BAS technology includes communication, networking and web services requirements for the systems which will be discussed in Chapter 11, but are clearly critical to unlocking the data from BAS sensors and energy meters to make them available for energy and analytics functionality. Finally it is critical to look at key drivers for future growth in the BAS industry, as they relate to this topic. This will be done as part of the technology discussions, yet there must be some general conclusions drawn about these trends and overall system requirements for the future.

The topic of BAS future systems is critical because ultimately these systems are, and will continue to be, restructured by technology migration from other industries and by the impact of IT based systems including middleware and web services. These IT based systems add a complex element to the BAS industry, yet as discussed, they are key to unlocking data, and the IT impact on BAS will continue pervasive and difficult to predict. Related to IT there are two critical points to make here: 1) BAS hardware components are in a dramatic state of flux and there is rapid evolution toward generic devices, and 2) HVAC and building functionality is exponentially more complex than any other IT service, such as data networking, email, etc. This means that there is no logical path to generic BAS systems with interchangeable components, and software is the key issue because of the substantial content knowledge required to develop effective BAS control strategies. As a result, the reader is cautioned to focus equal or more emphasis on finding the right company to do that work as on the technology itself. To provide a general BAS understanding, this chapter will outline the types of systems available, including the components and network infrastructure, along with a general overview of the features provided by these systems. This discussion combined with the chapters on networking/data communications and middleware should provide an excellent foundation for understanding the complexity of deploying energy and analytics functionality in a $4^{Plex}$ *Energy* marketplace.

## HISTORICAL PERSPECTIVE

BAS have traditionally been referred to generically as control systems, and the primary focus was on heating, ventilation and air conditioning (HVAC). As early as the 1970s and 80s the topic of BAS was considered a catch-all phrase for HVAC controls, fire and life safety, as well as security, but the reference was limited. In fact, none of these systems were truly integrated for control, except in very specific examples that allow a sensed condition in one system to trigger a control sequence in another. Most important to this book is that the data in these systems were not shared; these control sequences were usually triggered by relay logic or a virtual/software point flag that caused a system to trigger the sequence. Security systems migrated to IT communication standards well before the new millennium, yet fire and life safety even at this writing are still proprietary systems that do not share data. In great part this is driven by life safety codes, but many in the industry might attribute it to a lack of willingness on the part of manufacturers in that industry to embrace change. With regard to HVAC, conventional control systems over the years have evolved from pneumatic and simplistic electro-mechanical systems to mini-computer based technologies and ultimately to the highly sophisticated microprocessor based control products and systems of today. The focus here is not to provide a discussion of older generation or non-computer based equipment, but there may be some value in researching those systems outside of this book. This is because it is sometimes easier to understand the complexity of HVAC control by breaking it down into individual elements such as temperature, pressure, equipment safety, occupant safety, etc. For overview of more recent BAS and direct digital control technology, there have been many texts published including *"Direct Digital Control, A Guide to Building Automation and Control Systems"* by John J. McGowan, CEM.

The primary intent of this chapter is provide some context for not only understanding the technology, but also the industry and marketplace drivers that have led to, and are continuing to drive, significant developments. Some examples of past technology evolutions are the move from mini computers to microprocessors and distributed direct digital control (DDC) or energy management systems (EMS); legacy systems and proprietary protocols migrating to data communications standards like BACnet™; and then middleware (i.e., Tridium), convergence, web services, the cloud and now analytics. Many of these topics will be covered in upcoming chapters, but it is worth noting that the acronyms BAS, DDC and EMS have often been used interchangeably, and for all intents and purposes they are interchangeable. In this text, the term BAS is used, but the chapter title highlights that the BAS executes direct digital control and the acronym BA$^{DDC}$ will be used as well. The logic for using BAS as the technology descriptor is that DDC initially was a term that described where the control logic resided for a particular application. The advancement that made DDC possible was development of microprocessor-based controllers, which we now consider to be a requisite part of the BAS distributed computer architectures. These developments enabled migration of the control logic to the individual pieces of equipment, thus enhancing the granularity of control (temperature, pressure, safeties, etc.) and improving control reliability. The final point regarding system naming is that all industries are driven by markets, and when energy captures the customer's consciousness, a BAS may be called an EMS. That point provides a good segue into discussing BAS markets, prior to more discussion of technology, architecture components and features of BAS executing distributed direct digital control (DDC).

## BAS MARKET

The topic of markets is important to this discussion because BAS technology has traditionally responded to "market pull." Therefore as noted, when energy efficiency is a hot topic BAS are often called EMS, but also the technology evolves to provide more solutions for that market need. At the same time when customers are concerned about stand-alone reliability and resiliency of critical applications, distributed DDC may be more important to emphasize and be the driver for new solutions. Dashboards along with energy and analytics tools are highlighted here as major market trends that may impact system naming as well, but these systems have been compartmentalized and remain independent to date. Of equal importance is that market needs, as well as technology advancements, also open up new trends that go beyond the technology itself to create new business models. These models are often the result of leveraging these technologies with "know-how" to create value by offering such services as energy savings strategies, indoor air quality, etc. All of these developments will inform and dictate how these technologies and offerings evolve in this space. Consider how the advent of distributed DDC caused a business, which was dominated by three manufacturers with vertically integrated businesses selling direct and through distribution channels, to expand into a robust market with 100 or more manufacturers selling through multiple channels. The same was true when open systems and then middleware spawned a new breed of specialty contractors called a "system integrator." Non-technology events like climate-change and the economic downturn that started in 2008, resulted in the resurgence of a retrofit market, driven by energy management and by building improvements. The savvy business people among us are trying to determine what will be the next major market driver that shakes the industry. Here is the answer: It is called *Energy and Analytics for Buildings.*

Given that introduction, it is possible to turn the clock back for a brief history of how the BAS market has evolved. The first direct digital control product was deployed at an oil refinery in Port Arthur, Texas, around 1959. Under this technology section, the book will cover architecture and device functionality in more detail with more discussion of the systems themselves and how they operate. With that project in 1959, the first of these devices providing inputs (i.e., sensors) and outputs (i.e., control relays) (I/O) were deployed at a local process. Computers were applied to process automation and heating, ventilation and air conditioning control (HVAC) for buildings in the 1960s in the form of both direct digital control (DDC), and the setpoint control market has evolved in recent years. During this time however, the primary approach was to apply minicomputers or central processing units (CPUs) for building automation. These minicomputers held all of the control algorithms and tended to execute control through conversion panels that took the digital signal through an analog interface to the pneumatic systems that were prevalent in larger more complex buildings. In spite of the limitations of this approach, significant growth occurred through the late 70s as a result of the energy shocks of 1973 and 1977. This growth further accelerated with the development of microprocessors that enabled distributed direct digital control systems. These systems were capable of complex HVAC application, as well as digital control for energy management and also applied new energy saving features for optimizing equipment operation, offsetting electrical demand, and scheduling equipment to be shut down when not necessary.

The demand for BAS has grown and continued to evolve due to a variety of market trends. The demand for technology to save energy waned in the 1980s before the green movement caused a major market resurgence later in the 1990s and beyond. During the time when interest in energy efficiency waned however, the implementation of these systems over prior decades led to an awareness of the many benefits available through automation control. Building owners saw that beyond energy cost reductions there were real benefits to better control. These benefits included longer equipment life, more satisfied tenants, and a host of new features available due to expanded building information. The communication features available through these systems allowed for managers to track their buildings more effectively and respond quickly and intelligently to problems. Intelligent response was a function of BAS communication and the ability to remotely diagnose problems, which was the precursor to development of current analytics technology. These features were extremely desirable in all types of buildings including commercial real estate. In talking with property managers, and reviewing the results of surveys by the building operators and managers associations, it is clear that tenants often begin looking for another office space when there are space comfort problems. At the same time, many tenants choose to office in "green buildings" as well. This topic has begun heating up since the early 2000s when New York, Washington DC, the State of California and others began passing laws requiring building owners to track and report on building performance. There is the belief that this "CarFax" type of reporting for buildings will make the value of energy hog buildings drop over time and create another incentive for upgrades. The BAS executing DDC market therefore continues to be driven by savings in energy but more importantly dollars, and further by enhanced control and information

Due to the market drivers listed and others in certain geographic regions, the BAS market continued to grow in the wake a limited interest in energy during the late 1980s and early 1990s. Eventually these factors fueled deployment of BAS in every new building and caused further market growth from an industry of a few hundred million in the early 1980s to a projected global market of $49.5 billion in 2018. Getting to that size was due to a continued growth trend during the years of strong construction markets in the U.S. and the Pacific Rim in the early 1990s, followed by the Middle East building booms in such areas as Dubai, Abu Dhabi in the United Emirates and elsewhere. This boom continued until the global economic crises at the end of the new millennium's first decade. The BAS became an essential component of almost all new commercial and institutional construction during those years, and that was supplemented by the growing green intelligent building movement. Emphasis on LEED and Energy Star Portfolio management scoring that could be enhance with automation, enhanced the market acceptance. Interestingly the BAS market maintained some momentum even during the downturn due to the opportunity for retrofits in existing buildings.

It should be clear at this point that the current state of BAS is made up of systems that execute direct digital control (DDC) and enable communication between components via standard data communication protocols like BACnet™. Since DDC has been highlighted, it is worthwhile to define the term. The American Society of Heating Refrigeration and Air Conditioning Engineers (ASHRAE) defines DDC as a closed loop control process implemented by a digital computer. Closed loop control, the reader may know, simply means that a condition is controlled by sensing the status of that condition, taking control actions to ensure that the condition remains in the desired range, and then monitoring that condition to evaluate whether the control action was successful. A simple example is proportional zone temperature control. The zone temperature is sensed and compared to a setpoint. If the temperature is not at setpoint, a control action is taken to add heat or cooling to the zone, and then the temperature is sensed again. A requirement to accomplish this by digital computer does not impose any particular complexity, because today even the simplest residential thermostat is likely to have electronic or digital circuitry. The author therefore expands that definition to require that the digital control be provided with a microprocessor-based device that implements a sophisticated control loop, or sequence, and is capable of communication over a standard or "open systems" network.

With the advent of web services, the market has been preoccupied with the notion of controlling devices from the internet. However this is not a likely scenario for equipment or specific building device control. The reasons quite simply go back to that first demonstration in Texas: providing control intelligence at the device level provides for the fastest and most effective processing of control algorithms. Waiting for information to be communicated over the internet for a control action will undoubtedly take longer and cause a controlled condition to float or change before the action is completed. Not all control loops require fast response, but many such as humidity or pressure control can and do require speed. As a result, web-based applications are far more appropriate for such functions as analytics, which are operating on data that include historical and current performance data to develop decision-making information. These technical progressions in this industry have been spurred by ongoing demand for energy management, but also by a variety of drivers including open systems to ease integration of legacy technologies, integration of numerous types of systems like computerized maintenance management and security, as well as standards requirements like Energy Star and LEED to name a few. The future for this market will continue to be impacted by many of the factors outlined here. Equally important is the fact that the Big Data for energy and Buildings movement will be a large driver because these systems are essential to gathering current and accurate data about building operations.

## BAS TECHNOLOGY

BAS, and more specifically $BA^{DDC}$, technology will be discussed within the context of control architecture. Architecture in this case refers to how $BA^{DDC}$ devices are physically wired together, but also to the notion of how they interact with one another. The BAS enterprise is the highest level of architecture and is enabled with the greatest set of capabilities. As this discussion unfolds, the reader will begin to grasp system interactions more clearly. With the advent of the internet and web-based applications the architecture can be viewed as local (in the building) or global (connected via the internet). The goal here is to ensure that the reader has a good basic understanding of BAS architecture at the local, building or campus level. Architecture, as it relates to all computer based systems, will be defined in more detail in Chapter 11, but for this discussion it may be viewed as a group of related control devices and a common communication network, such as BACnet™. Again the concept of a local area network (LAN) for building automation systems (BAS) will be discussed further in Chapter 11, but a definition of the LAN is still necessary. One comment prior to that definition however, is that this book treats the building automation system as being made up of control system components, which are referred to as controllers. The emphasis here is on energy and analytics, which again require data from the controllers within the BAS. The BAS focuses on building-wide control, and the integration of that control via many distributed devices or controllers. Distributed control devices rely by necessity on networking, within an architecture, which will be discussed in the next chapter. In addition, energy and analytics can, of course, complete equipment performance evaluations that may be used to enhance control functionality. Solutions providing such services are being introduced to the marketplace on a continuous basis.

Returning to the topic of architecture and BAS networks, a computer local area network (LAN) as defined by Thomas W. Madron "covers a limited geographic area…," where every "node on the network can communicate with every other node, and… requires no central processor."* A node is any device on the network, and for a BAS this is typically an intelligent microprocessor based controller. The control industry adds more complexity to the idea of a LAN. Unlike computer LANs, the requirement for the BAS is to provide control at an individual piece of equipment, and to also provide coordinated control for multiple pieces of equipment at multiple levels of architecture within the system. These concepts will be the focus of this chapter to set the stage for more detailed discussions of BAS protocol and networking, as a requirement for energy and analytics functionality.

The overall control architecture will be discussed in terms of the individual components. To consider the control architecture there is merit in briefly discussing a simple control application. A device within a simple BAS executing distributed DDC architecture provides control for HVAC equipment and ancillary loads such as lighting. Accomplishing this requires that distributed controllers be applied at each piece of equipment to provide application specific control. These controllers must be able to implement closed loop control via monitoring of status conditions and execution of control actions. At the same time it is also necessary to integrate overall building control provided by these devices for such functions as optimizing the building wake-up or start time. There may also be a requirement for virtual points that come from outside the control network via the internet, such as a demand response signal. These types of features for BAS systems executing DDC will be discussed in more detail throughout the book.

---

*Thomas W. Madron, *Local Area Networks, The Second Generation*, John Wiley & Sons, Inc., New York, N.Y., 1988, p. 3

This chapter provides the user with a basic understanding of the control architecture and technology necessary to implement a building automation system executing DDC. These systems are made possible through application of powerful microprocessor based controllers with distributed intelligence capable of equipment control, standard data communication networks such as BACnet™ and internet functionality for remote communication with energy systems, dashboards, analytics and much more.

In a typical BAS, four levels of architecture are defined: sensor/ actuator, distributed controller, enterprise supervisor and dashboard. The BAS interface function is rapidly migrating from the proprietary operator interface (POI) to dashboards. All of these levels of the BAS architecture have undergone change in the last two decades and therefore each will be discussed.

The sensor and actuator technology level has traditionally been hardwired, but that is changing too. Traditionally sensors and actuators only used electrical signals rather than networking technology. Industry standard signals like 4 to 20 milliamps that are wired to a controller are most common. Controllers must then use analog to digital, or A to D, conversion technology to convert the signal to an engineering unit for use with the control sequence. For the purposes of this book, we will not delve into this technology except to note that sensor based local area networks are not far from deployment with lower cost and more compact nano-circuitry. The factors that will determine how quickly these devices achieve market acceptance will have to do with enhanced features, game changers like reliable wireless networking, cost of the technology, as well as the installation and need for the data. Interestingly, energy and analytics could be one of the factors that drives this evolution, because not all of the data are currently installed, and that will require further deployment that can be costly.

At the upper level of architecture the dashboard interface is a critical system component, yet it is ancillary to control. This level will not be discussed here, because it is covered as part of the enterprise supervisor section, an entire chapter has been devoted to the topic, and it will be treated throughout the book.

An interesting note regarding building automation systems is that there may be sublevels within the architecture. The most common implementation of distributed systems is to have both a zone level controller, tailored to simple applications like the VAV terminal unit, as well as an equipment level, or general purpose, controller for larger pieces of equipment like the VAV air handler. These two types of controllers will have different requirements and designs to accommodate the given application, yet they both fit within the level of a distributed controller. Therefore this discussion will focus on the control specific components of the system:

- The enterprise supervisor,

- BA$^{DDC}$ equipment level controller, and

- BA$^{DDC}$ zone level controller.

A typical building automation system executing DDC architecture will include an enterprise supervisor at the highest level and two levels of distributed controllers. The discussions of each of these levels of control will consider specific hardware and sequence of operations for that particular component. It is also important to cover the concept of building-wide or integrated control via the enterprise supervisor. Finally, this technology is extremely dynamic, so a general overview of trends will be provided for each major component of the BAS, as well as overall trends for BAS as a whole.

**Enterprise Supervisor**
The enterprise supervisor was selected for discussion first under this BAS technology section because it is responsible for overall building coordination. Key topics to this discussion are hardware requirements, building coordination which provides a sequence of operations for the entire system, and "analytics impact." Energy and analytics impact relates this BAS topic to the broader topic of the book, and will emphasize why BAS is so critical to accessing data and making it available to big data systems via the internet, etc. Throughout this chapter, each topic also addresses market and technology trends that can be expected to impact each level of control.

*Hardware*
Enterprise supervisor hardware varies greatly. In the past, approaches have been taken from custom computer based systems to dedicated personal computers (PCs), as well as to develop custom microprocessor-based equipment. It has become an increasingly common design concept to develop "platform" microprocessor-based devices.

Such devices package the microprocessor with an operating system, non-volatile (read-only memory) and volatile (random-access memory) storage, software programming and configuration capability and capabilities for data communications networking with other BA$^{DDC}$ devices and, as appropriate, with the internet. In some cases the enterprise supervisor will include on board componentry (inputs and outputs or I/O) to sense external conditions and execute control action, but this is less common. This custom microprocessor approach is more common today in great part because of standardization, cost reduction and hardening of the technology to provide optimal performance. The custom microprocessor-based enterprise supervisor (ES) must be viewed primarily as a vehicle for managing the network, integrating building-wide control among all the distributed controllers and providing central communication functionality including the BAS device network, as well as internet access for wide-ranging capabilities. For clarification purposes, the ES is managing a device network of building automation direct digital controllers (BA$^{DDCs}$). By their very nature, these BA$^{DDCs}$ execute control via distributed processing at the device, or at the individual piece of equipment, level. These concepts will be expanded significantly under sequence of operations, yet they are mentioned here to indicate the hardware complexity required at the enterprise supervisor level. In spite of the large number of variations on manufacturer approaches to development of this hardware, the author will describe a somewhat typical hardware example. A final important point is that the term "enterprise" was chosen due to the focus on a particular building or campus environment. Later in this book "middleware" and web-based systems will be discussed that interface with the enterprise and may introduce new virtual data etc., but the distinction here is that this enterprise supervisor is managing network communication and control execution for the architecture of BA$^{DDCs}$ making up a local system.

As noted, the enterprise supervisor (ES) is typically focused on providing device networking, management of control execution and internet access functions. Therefore the ES may or may not have on-board inputs and outputs (I/O) to physically sense information and directly execute control. In some cases these devices will have a limited number of I/O, but the trend is certainly toward use of system-wide data in the form of virtual, or software, points. A virtual point is a piece of information that physically resides elsewhere in the system, or may come from an external location via the internet, but is held in a "data field" at the enterprise supervisor, for communication to other distributed controllers, or BA$^{DDCs}$. Perhaps the best example of this is outside air, which is needed by most BA$^{DDCs}$, especially HVAC. An example of an internet based virtual point might be a demand response event notification, which is shared with all BA$^{DDCs}$ so that they may execute electrical load shed. Along with physical points, which may be connected to any particular BA$^{DDC}$ device, both types of points can be shared via the network with all other controllers. These data-sharing functions will be discussed as part of control integration under sequence of operations. From a hardware perspective this means that the enterprise supervisor becomes the focal point for networking and communication within the system and with the outside world. Communication with the outside world is provided through an internet protocol (IP) address, and internet access, allowing access to all system data from an authorized external "dashboard" or interface. Under the networking and communication chapter the types of IP addresses, and the challenges associated with ensuring that valid addresses are maintained. Of course, there are still some older legacy systems that used modems and telephone lines, but these are rapidly vanishing. As noted, the enterprise supervisor provides overall network management functions, though in many cases these systems are self-policing. Self-policing in this context means that each network member, or BA$^{DDC}$, has responsibility for ensuring communication and sequences that are automatically enabled when a failure occurs. There will be much more discussion of these concepts later in this book.

The enterprise supervisor hardware described here is therefore a powerful microprocessor based system. This system is focused primarily on the two levels of system communication: networking and remote interface. The enterprise supervisor must also coordinate building-wide control functions and availability of system-wide data sharing, which will be discussed in a moment. These functions are typically implemented in one piece of hardware, however there may be other functions, or applications, required for a BAS such as life safety, fire and video surveillance/access control security systems. For purposes of this discussion, the assumption will be made that these functions are provided via separate "application-specific" enterprise supervisors. In fact there may be a system-wide requirement to integrate control between these various application areas, yet this chapter will focus on building control primarily for energy-consuming equipment. As noted previously though, it is important that some energy and analytics processing requires data from other sources to develop metrics and key performance indicators (KPIs), yet again the focus here is on buildings and energy. For the most part it will be assumed that the dashboard is only concerned with monitoring and access of these other functional systems. In summary therefore, this hardware overview defines a device with central focus on enterprise communication and networking and little or no physical I/O capability. Next it is necessary to consider building-wide coordination of control.

### Building-wide Coordination

Building-wide coordination with a BA$^{DDC}$ distributed system will be viewed in three key areas:

- control integration,

- building-wide monitoring and short-term data history, and

- remote communication with dashboards and with energy and analytics tools.

With the introduction of BAS executing DDC (BA$^{DDC}$) a number of new issues were raised around the accomplishment of control integration. Again these are distributed processing devices by nature, so the focus at the device level is on a specific piece of equipment or control application. The author uses the term integration because of the need to accomplish control via coordinating, or more specifically integrating, the functions of multiple microprocessors to provide control for those applications. Integrating control is important because it means the accomplishment of complex strategies that are implemented in part by more than one device. There are many functions that fit into this category and examples include morning warm-up in a VAV system, demand response or optimal start/stop for the building. On a building-wide level, the BAS must monitor functions at each device, share information about those functions via the network, and then integrate the control sequences carried out by all of the devices.

There are many design approaches taken with BA$^{DDC}$, due to the power of distributed processing providing by these systems. For simplicity's sake this discussion will take one of the more common approaches, enterprise supervisor integration, and discuss it further. In this approach the enterprise supervisor provides building-wide functions by monitoring building conditions or status, and based upon that data executes DDC control sequences. For example the enterprise supervisor will monitor building power consumption for the network, and based upon demand consumption and a setpoint will initiate demand shed. At the same time the enterprise supervisor will monitor the internet for a demand response (DR) event signal. The sequences executed under these two scenarios may be the same or different depending on the building and its kW electric demand profile. In either case the actual control approach focuses on control of loads that can be shed (turned off) through integrated control sequences implemented by BA$^{DDCs}$ throughout the building. These BA$^{DDCs}$ must physically disable loads to avoid exceeding the setpoint (or for DR the kW drop contracted with a utility) because the enterprise supervisor may not have access to actual controlled devices that can be shed. In both of these cases it also becomes critical to understand the building and use the BAS to monitor conditions, so that critical conditions do not "spin out of control" during the shed period. Many of the largest loads (most kW demand) that are easily accessible for demand limiting and DR, are HVAC. For HVAC to "spin out of control" would mean that a building exceeds the acceptable comfort conditions and becomes an unproductive environment for the work being done in that building. The real power of analytics is that operators can develop analysis routines that compare a series of variables regarding indoor comfort, outside temperature, equipment loading, etc. With that knowledge it is possible to fine-tune the building by creating sequences that can avoid unintended consequences from trial and error approaches that may result in negative outcomes.

The key to control integration is the ability of the enterprise supervisor to either "request" the implementation of local control sequences, or to override local control and force a specific sequence. In either case the enterprise supervisor must track the actions of BA$^{DDCs}$ and be capable of taking appropriate steps if the sequence is not implemented. There are many schools of thought as to the best possible approach for accomplishing these types of functions. For this discussion however, it is only important for the reader to note that building-wide control sequences require integrated action by multiple BA$^{DDCs}$ to be achieved.

The final topic under control integration will be system data. The term virtual point was used earlier to identify information from a sensor, etc. that is not physically connected to the ES, but the concept of system data provides even broader context. In essence the enterprise supervisor serves as the vehicle for ensuring that key information is available throughout the network. Within the BAS networks, there are two types of system data: global and shared. Global data are information needed by many or all of the controllers in the same format. There are two types of global data as well, Global$^L$, for data from "local" BA$^{D-DCs}$, and Global$^I$, for data from "internet" sources via the ES's IP connection. Good examples of Global$^L$ data are outside air temperature, time and enthalpy. These data are provided by one controller which has a physical point for monitoring, though for system integrity purposes backup locations may be provided. The data from that hardware point are then "broadcast" to the network and available to all BA$^{DDCs}$. A good example of Global$^I$ data is a DR event signal. These data are provided by the ES from the internet and the data are then "broadcast" to the network and available to all BA$^{DDCs}$.

The second type of system information is shared data. This is a much more custom approach to using system data for control. A simple example that will also be discussed later is temperature reset. BA$^{DDC}$ A is responsible for maintaining a discharge air setpoint, but that setpoint may be adjusted based upon an indication of the load. BA$^{DDC}$ B is monitoring space temperature for local control, but this is also a good indication of the load. The shared data function provides a vehicle for controller A to get access to the space temperature with the same frequency as if it were physically connected to that device. This requires that a specific point be accessed by the enterprise supervisor with regular frequency comparable to the need for control, and that the data be transmitted to controller A. These types of functions are critical to building-wide control.

Another building-wide function of the enterprise supervisor is to serve as a vehicle for monitoring and short-term operating history data. Monitoring functions will be discussed under the dashboard portion of this section, yet the most important point is that a "window" to the network must exist. This window must allow for communication with all BA$^{DDCs}$ via an IP address for the local building or campus network, providing internet access. Through that window a variety of monitoring and alarming features must be provided as will be discussed in a moment. Also of critical importance is short-term historical system operating information.

Historical system operating information is BAS data, and requirements for these data have been expanded dramatically. In the past these data were used primarily for alarms to identify issues that required immediate attention, and for "trend" data that could be used for trouble-shooting and diagnostics. In this new "energy and analytics era," the requirements have expanded. For purposes of this discussion, consider one application "commissioning." Commissioning requires much more data for the equipment to control or troubleshoot, and it must have those data for longer time periods, and possibly at a more granular level to achieve its purpose. Prior to the energy and analytics era, the enterprise supervisor served as a central repository for data files containing trend data on various system points. This was generally the most efficient approach to ensure that the overall cost of the system is not detrimentally impacted by the need for data storage at every BA$^{DDC}$ device on the network. To reduce the cost burden on the overall system for storing data, the general approach was to charge one network device, the enterprise supervisor, with responsibility for history. This required the enterprise supervisor to access data from all BA$^{DDCs}$, and to provide data storage for later retrieval. That solution would be cost prohibitive in the energy and analytics era due to the volume of storage that would be required for commissioning alone, but add enterprise energy management (EEM) functions, plus other applications, and the system would be unsustainable. The good news is that there is now an easy solution. IP access and the evolution of server farms to provide large volumes of "cloud" storage make it possible for the ES to store data locally for a short period of time and then upload it to the cloud at specified time intervals or even stream it. From a marketing perspective, this energy and analytics era functionality points to the reason why analytics companies are typically agile technology companies rather than large BAS manufacturers. Without question many BAS manufacturers are integrating analytics into the BAS, in much the same way as an "APP" is added to the smart phone, but that may not be the only path to market. Because of the large requirement for data storage and the need for processing and analysis of the large dataset that now resides in the cloud, not at the ES, other solution providers are stepping up to perform this analytics function. The reader is encouraged to pay special attention to the technology evolution to ensure that specific requirements for any building or applications can be met by a particular architecture implementation.

The final enterprise supervisor function is to allow for remote communication "interface" via BAS software or dashboard. For simplicity purposes, all of the variations on software interface technology will be referred to generically as "dashboard." Traditionally there were three key functions required of an interface: programming, monitoring and alarming. Given the evolution of post-BACnet™ BA$^{DDC}$, there are currently two functions: programming and monitoring/alarming. Yet this aspect of the BAS industry is in a significant state of flux. Beginning with programming, this function requires that each BA$^{DDC}$ network member including the enterprise supervisor may be programmed and configured using what is now called a workbench tool. Programming and integration are covered here briefly for information. Every BAS device including the ES and BA$^{DDC}$s must be configured with identifying information. Configuration happens only once, and it includes such information as an IP or network address and essential data to identify that device on the network, and to establish specific characteristics like whether I/O is connected. For BA$^{DDC}$s in particular, programming will include a sequence of operations that it is to execute and I/O devices necessary to execute that sequence. Programming will also include a variety of parameters such as control setpoints and alarm limits. It bears noting that when the building automation and control network (BACnet™) standard was promulgated, the American Society of Heating Refrigeration and Air Conditioning Engineers (ASHRAE) standards committee established that this programming function was not to be part of the standard. The reason was simple, the BAS industry is made up of companies who believe that their sustainable competitive advantage is innovation and creativity in developing and enhancing control sequences. The BACnet committee elected not to try to standardize control sequences, but rather to respond to the industry demand, which was to provide for a standardized networking and communication protocol that would allow operators and engineers to avoid being locked in to specific systems. This decision resulted in the birth of "integration" as a specialty, because companies had to develop specialized knowledge that allowed them to coordinate the operations of multiple BAS systems within a single building or campus. BACnet provided the basis for accomplishing this through the second category of interface: monitoring/alarming.

Monitoring/alarming is truly more than that for BACnet based systems, as for those that use any other network communication standard. Dashboard interfaces take their cue from BACnet by enabling monitoring, operations and alarming, and included in those categories is "data management" for analytics systems that leverage cloud computing. As will be discussed in the next chapter, BACnet enables standard-based interface to all BASs that subscribe to the standard, and equally important it enables the operator to change a number of programming parameters, such as setpoints, schedules and alarm limits. Dashboards have access to current status on all controller points, as well as access to system trend and alarm history data. Dashboard alarming is a critical feature of any BAS, and can report on critical conditions that may be defined as well as default alarms, such as sensor or controller failure. The system designer may then note whether such alarms should be logged for availability when the system is polled or should initiate notification over the IP network. IP network notification uses the internet capability to initiate communication with one or more predefined dashboards and send the alarm data. These functions will be also discussed in the networking and integration chapters.

As noted throughout this section there are a number of market trends in play at the enterprise supervisor level. This section will not repeat these trend discussions, but there is merit in a general statement about one of the trends that has not been mentioned, the growing interest in BAS from the IT industry. At the Realcomm/IBcon conference (www.realcomm.com) in 2014, the author witnessed a keynote panel with Microsoft, Google, Cisco and Intel discussing their respective positions on the future of BAS. The significant interest in green buildings, smart/connected buildings and sustainability in general, are just a few of the possible reasons for this interest, but beyond that is the obvious opportunity for these companies to create more demand for their core products. The reader is encouraged to watch these trends and see how they unfold. However, the balance of this chapter focuses on content that shows the true complexity of BAS applications. That complexity means that IT companies have a big lift to effectively match their technology expertise with building operations.

**Energy and Analytics Impact**

Throughout the book so far there has been a good deal of discussion about the criticality of data access and data formatting to energy and analytics tools. One of the greatest barriers to entry for analytics over time has been the cost of instrumentation to provide data for analysis. Perhaps the most important impact of BAS on energy and analytics is that these systems already have access to extensive amounts of data. Even better, in many cases it is just the equipment condition, operations, building environment and energy metering data that are essential for performance of meaningful analysis. The second critical point regarding enterprise supervisors is that these devices typically conform to data communications and networking standards. Now that does not mean the data themselves are in a standard format, but it makes accessing the data significantly easier, and through BAS IP addresses it is possible to push large amounts of data to the cloud or other server. The issue at the cloud is that metadata and tagging/naming standards do not exist, so the analytics engine can make immediate use of the data without the need to massage and convert it into a standard format. Project Haystack (see the chapter on this topic) is about making this process easier and more seamless by creating a data model. The data model create a framework to standardize data tagging, etc. Performing an extra task to convert data from many different BASs (typically called mapping) can be cumbersome and time consuming, but it is still dramatically more cost effective than installing an extensive instrumentation network. In the end, there will be many benefits to Haystack, and other data standardization efforts, but among the principle benefits will be faster technology deployment and accelerated product development of new energy and analytics tools, features and functions. As the dataset of information points from BAS, as well as all other systems that interact with the building, expands and become richer, it will be possible to drive energy and analytics functionality that we cannot conceive today.

## BA$^{DDC}$ TECHNOLOGY FOR EQUIPMENT AND GENERAL PURPOSE APPLICATIONS

The enterprise supervisor discussion focused on a top-down view of BA$^{DDC}$ technology, yet equipment control is what actually distinguishes these systems from other types of computer-based or IT applications. This section will provide a generic overview of applications that addresses the larger and more complex equipment in a building, as well as a variety of other devices and systems such as lighting. There are many features and capabilities available with BA$^{DDC}$s for more complex applications. Microprocessor based control is ubiquitous with large heating, ventilating and air conditioning (HVAC) equipment, often installed at the factory by the HVAC manufacturer, but it is also common for general purpose control. The term equipment level is derived from the use of these controllers with such equipment as centrifugal chillers and air handlers. General purpose control uses the same hardware, with slightly different control sequences, for such applications as start/stop of non-HVAC loads or on/off control of lighting and other electrical equipment. BA$^{DDC}$ control with equipment level and general purpose controllers is highly cost effective due to application complexity. In these applications there are also greater requirements for control integration due to the distributed processing nature of BA$^{DDCs}$. Also the equipment cost has an order of magnitude that can more easily support the cost of BA$^{DDC}$ for multiple reasons. As HVAC manufacturers have gained market share and joined the ranks of market leaders, these companies have utilized BA$^{DDC}$ to replace thermostats and integrated equipment controllers for static pressure, air volume etc., but also for equipment safety. It has been possible for equipment manufacturers to program a wide range of alarms, diagnostics, troubleshooting routines and protective sequences, that in the past required independent devices. The addition of such technology also makes it possible for these manufacturers to consider offering new services to the market like commissioning or energy analytics, but more on that later.

The enterprise supervisor has been discussed as a building-wide integrator of control functions. Yet the benefits of BA$^{DDC}$ are truly realized when intelligent devices are deployed locally at a particular control application. Equipment/general purpose controllers form one of the two levels of BA$^{DDC}$ devices noted. They provide full local control requirements, and integrate with both the enterprise supervisor and zone level controllers to provide building-wide functions. The topics which will be covered under this heading include the following and as appropriate the sections will also address relevant development trends:

- Equipment level hardware,

- Sequence of operation programming,

- Control integration,

- Networking, and

- Energy and analytics impact.

### Equipment Level Control (ELC) Hardware

As with the enterprise supervisor BA$^{DDC}$ ELCs are typically "platform" microprocessor-based devices. These devices encompass the microprocessor, as well as an operating system, non-volatile (read-only memory) and volatile (random-access memory) storage, on board componentry to sense external conditions and execute control action, software programming and configuration capability and capabilities for data communications networking with other BA$^{DDC}$ devices and, as appropriate, with the internet. Among these characteristics, perhaps the most distinguishing feature of equipment level hardware is the number of on-board inputs and outputs (I/O), which are also referred to as "points." Devices applied at this level provide both analog and digital inputs and outputs, as well as network access to virtual points to introduce new variable information for control sequences. Input/output (I/O) requirements are generally more demanding at this level due to the expanded applications. There is generally a demand for at least eight of each point type, analog and digital inputs and outputs. Due to the fact that these controllers may be applied as general purpose devices or application specific controllers, it is not uncommon for controllers to be expanded up to as many as 64 or more total points. This may be done through add-on modules, cards, etc. It is also necessary because most equipment applications require control through digital outputs and analog or pulsed/modulating outputs, as well as monitoring of both analog and digital values. Analog values are generally used to monitor temperature or pressure, while digital points provide contact closures typically indicating an alarm. Device outputs then respond to the commands of the controller based upon the appropriate input value and the sequence of operations. These outputs will either be enabled or disabled, or may be modulated to a commanded position.

As noted, it is common for manufacturers to develop a platform device for equipment/general purpose hardware and use it with all applications. The platform will provide a base input/output capacity, and in many cases allow expansion through any of several techniques. Application of the controller is accomplished by changing software in the device. Software issues related to this approach will be discussed further under sequence of operations. Hardware design issues to be resolved also include the requirement for a full line of ancillary component products. These products include the:

• controller,

• sensors, transducers, other input devices and

• electrical relays or contactors, motors, actuators, transducers and other output devices to enable control points.

A key related issue with these products is that, in the past, many manufacturers would fix the ancillary components (sensors, transducers, etc.) that could be used with a device or allow flexible application of a wide variety of industry standard equipment. This is less common today because most integrators demanded the flexibility of choosing sensors, etc. Another benefit of using generic ancillary components is that the cost of system retrofit may be greatly reduced because existing field devices can be reused. The intent of this section is not to provide a detailed discourse on ancillary devices, but it bears noting that the "Internet of Things," discussed later in the book, and the pervasive deployment of an infinite number of ancillary components has made it cost effective to begin networking these devices. This topic will not be discussed in great detail here, but the economies of scale associated with the large volume of ancillary components being applied, will lead to a new level of architecture. This level of architecture may be called instrumentation, ancillary gear or some new name, but the expectation is that networking cost declines combined with such advancements as nanotechnology will produce a great deal of advancement in this area.

There are two additional points that should also be mentioned under the category of hardware: controller size and ambient operating requirements. Controller size is typically not an issue with general purpose devices because they are often mounted in a mechanical room, above a ceiling, etc. The assumption is that these devices can be acquired with enclosures that meet standards agency approvals and local codes. With equipment controllers as noted, manufacturers generally pre-integrate BA$^{DDC}$ devices within the control panel of HVAC equipment. This imposes a size constraint in smaller units. Another topic under hardware is the absolute requirement for equipment-level control devices to be able to operate under outdoor temperature conditions. A good example of this is a VAV rooftop air handler. It is typical for this equipment to be provided in a fully self-contained manner. A single enclosure containing the HVAC equipment and controls is provided in one package that is mounted on the roof. The package will be exposed to the full extreme of outdoor temperature and humidity, and may be shipped to any location in the world, further adding to the potential extremes. In closing, it is also important to mention "middleware" and integrated solutions. To a great degree, the proliferation of options available for BA$^{DDC}$ systems has resulted in deployment of multiple different devices that must interoperate. The sophistication of system integrators combined with the diverse number of devices that can execute control in a "multi-vendor BACnet" environment has made this possible. Middleware and integration will be discussed further in this book, the primary reason to mention it here is that the architecture of BA$^{DDC}$ systems has morphed over the years, and it is not possible to make simplistic statements about how equipment-level control is achieved. This is particularly true as BA$^{DDC}$ has become standard in nearly all commercial buildings, and BACnet specifications have been developed by engineers with the idea that HVAC equipment with pre-integrated controllers will operate seamlessly with technology that is site-applied by a control contractor. These systems can be very effective, but that is by no means guaranteed. This author has said more than once that the new construction "design, bid, build" market is hopelessly broken, and without a highly qualified integrator that is goaled to be a constructive member of the design and installation team, the end result is not always positive. The reader should not imply that this is an indictment of any particular participant in this design bid supply chain, it is rather a result of the predatory nature of the model.

*Sequence of Operations*

Control sequences truly define BA$^{DDC}$, and the requirements for equipment and applications targeted at this level are significant. The author has heard this concept referred to with a simple analogy. It may be related to a civic auditorium, for example Madison Square Garden. The basic BA$^{DDC}$ hardware platform outlined above makes it possible for that device to be applied in any number of applications, much like the functions in an auditorium. The sequence of operations is the specific program that tailors this BA$^{DDC}$ to the application. Correspondingly the marquee and associated interior modifications in seating, etc. within the auditorium tailor it to any given event. This section will briefly touch on the techniques that are used to accomplish that programming.

For the most part, BA$^{DDC}$ products on the market today use manufacturer-developed, "proprietary," programming tools. Generally these tools use one of two programming approaches: equation driven and library loop control parameter approach. This may come as a surprise for those readers who are not deeply immerse in the BA$^{DDC}$, but as has been mentioned in this text BACnet is not a standard for programming. BACnet is a communication network standard that is designed for use after the BA$^{DDC}$ is fully programmed, configured and commissioned. Again, the term "workbench" is becoming more common as a generic way to refer to this software, and there has been significant industry effort to develop "one size fits all" tools for this purpose. However, that notion is not widely embraced by the industry. Under this heading of workbench programming, it is important to note that the *"styles"* do vary. For example, equation driven programming may be done by coding using custom software created by the manufacturer, or ever an IT language such as "C." Another very common *style* is to perform this task using a Viseo™-type of block programming. More may be learned about Viseo at www.viseo.com. There is also industry interest in using the Building Information Modeling (BIM) tool for this purpose in the future. With this Viseo graphic *style* of programming each block is actually a logic block, represented graphically, that performs a mathematical function and the programmer builds sequences in a logic driven graphical manner. For example, a logic block may compare space temperature to a setpoint and based upon a calculation determine if heating or cooling should be turned on. This section will not discuss *styles* of programming and configuration further, but the reader is encourage to ensure that they understand the *style* that is used for any BAS they procure and that this software is provided as part of the installation contract.

Regardless of programming approach, a very important point for BA$^{DDC}$ manufacturers is that they believe their expertise in understanding the applications and creating effective control sequences is what differentiates them from competitors. Therefore they are not anxious to give up the credibility that this expertise provides them, nor are they willing to lose the intimate relationship with integrators, and building operators, that comes from selling, training and supporting the BAS. At the same time there are many other issues. Consider liability for example. If a third-party software package is used and something goes wrong with BA$^{DDC}$ operations, who is at fault?

This author, and others, have written extensively on those topics in other books, such as the one referenced earlier. So this section will describe these approaches in limited detail as a general introduction, and tailor the discussion to equipment-level controllers (ELCs). The complexity of control sequences required for equipment, in this case HVAC, is quite sophisticated because it requires control of multiple conditions, such as temperature, pressure, possible humidity or carbon dioxide, and often much more. These types of sophisticated control sequences will expand on simple proportional control. Proportional control is the most basic sequence of operations for HVAC. Earlier in this chapter, a definition was provided for direct digital control. In that definition, the idea of a feedback loop was introduced. Proportional control is the simplest form of such feedback loops and it is also sometimes referred to as "linear control" because the response to a change in the condition, i.e. temperature being sensed, is linear. Proportional control systems are more complex than on-off control, which could be like a bi-metallic domestic thermostat that simply turns or a heating unit when the temperature drops below the setpoint, or like the special-purpose devices that will be discussed later that turn a load on or off based upon a time schedule, etc. On-off control will work where the overall system has a relatively long response time, but will result in instability if the system being controlled has a rapid response time. Proportional control overcomes this by modulating the output to the controlling device, such as a continuously variable valve. Quite simply, "feedback" means that a condition, again i.e. temperature, is sensed and compared to a setpoint, i.e. thermostat temperature setting, and an action is taken to turn an HVAC unit on or off. The feedback loop continues to sense the condition and turns the unit on when heating or cooling is needed and off when the space is at the setpoint. There are a number of issues with simple proportional control, because often the unit will run longer than necessary and overshoot the setpoint, and the temperature will drop too much before restarting the unit. These issues affect comfort in a space, but they are also inefficient. These issues were among the factors that led to development of more complex control sequences that calculate proportional control, but also perform integral and derivative computations. In this space temperature example, the function of the integral would be to calculate the distance (# of degrees) that the space temperature is from setpoint and adjust control action to avoid wide swings around the setpoint. These swings are often caused by overshot of the setpoint and delays in responding to a call for conditioning which are common with proportional control. The derivative feature is very specialized, and is used with highly dynamic applications such as pressure control. Rapid changes in a condition require that the control is adjusted to react quickly. The result of a proportional, integral and derivative (PID) control loop is a condition that is accurately maintained at desired levels with very little deviation. Combining this highly accurate control with networking standards like BACnet is an essential feature of BA$^{DDC}$ for building control.

ELC control sequences will vary dramatically by type of equipment. These devices require complex and sophisticated control sequences tailored to the specific applications. One approach to solving this problem is to have the control designer, contractor or integrator write the sequence with an equation-driven language. This is done using "workbench" software and can address any anomalies or custom requirements, yet this task can also be time consuming. As the BAS industry has become more competitive and equipment, or integrator, selection is often made based on price alone, there has been a growing interest in development of library sequences. This is because library algorithms allow much quicker installation, startup and commissioning, but they do not allow for custom application. Another variation on this theme is to use a controller that allows flexible programming, and maintain a custom library. This allows the designer to start with a reasonably complete control sequence and edit that sequence to meet the new requirement.

The equation library approach works well for equipment level control because sequences may be maintained for chillers, boilers, air handlers, etc. Again many equipment level controllers come pre-integrated by the HVAC manufacturer and the sequences may be "locked in," meaning they cannot be modified without violating equipment warranty. So, the designer, or integrator, must determine how critical it is to modify the library, and whether these modifications will impact control reliability. If the ELC workbench software allows customer programming, it will of course be necessary to test and verify each new control sequence prior to field application. This testing can be extremely time consuming and cannot always verify every aspect of control operation. As a result the library loop control parameter approach was developed. One note is that some controllers, to a great degree zone devices to be discussed next, embed control sequences, like the pre-integrated controls mentioned above. However, it has also become common for workbench software to provide a library loop approach that allows for a variety of algorithms to be selected and downloaded to any control device.

Library loop control parameter type programming basically involves controllers that employ sequences that have been programmed, tested, verified and possibly hardcoded by the manufacturer. Hardcoded means that these sequences are held in a portion of memory that may not be accessed to review or change the sequence. As noted above, integrators often find it more convenient to access library loops on the workbench and download those into any device in the field, or during bench testing at a shop. Obviously hardcoded devices that are preprogrammed at the factory rather than allowing a workbench download are less desirable for some integrators because they must sacrifice a portion of the control flexibility available with free-form equations. Yet the engineering time and job support required to deal with writing and testing control loops is virtually eliminated. This is a strong selling point for such devices, and has led to a large number of products on the market that offer preprogrammed control sequences of various types. Another example of the library loop approach is with general purpose control. In this case general purpose is defined as simple control loops that provide either start/stop or modulating control. Examples might be time of day (TOD) scheduled sequencing of loads such as lighting and/or other electrical loads. These functions are among the first that are being completely standardized and embedded directly into BACnet. Yet these remain perfect examples of control parameter applications, because it would not be an efficient use of time to write equation programs for every TOD schedule. Another example would be a simple proportional type temperature loop. Such applications might be a three-way valve on a chilled water supply loop that is modulated to maintain a constant loop temperature. Again custom programming is not likely to be cost effective for such an application. The intent of this section is not to position one approach as desirable over the other, rather to identify the technology used.

To be more specific about the technology is difficult in this section, however there is another important concept. That concept concerns the complex nature of BA$^{DDC}$ equipment level control. To illustrate this point, consider a variable air volume (VAV) air handler. The control for this piece of equipment must provide discharge air temperature control by sensing temperature and modulating an outside air damper for ventilation, and possibly free cooling, along with heating and cooling equipment. Often these units are cooling only, however it is not uncommon for air handlers to provide heating as well. This means that the BA$^{DDC}$ requires one control loop to make a control decision on the mode of operation, heat or cool. Two additional control loops, one for heat and one for cool, then determine the amount of heating or cooling to provide. The control action is based on staging of independent capacity, but may also include an outdoor air economizer sequence, as noted. This fourth control sequence, the economizer, must consider another input, outside air temperature. The economizer may be required to consider enthalpy, rather than temperature alone, to make a decision about modulating the damper to provide free cooling. A fifth control loop that is common with these systems is temperature reset which will look at one of several temperature conditions, i.e. space temperature, and reset the temperature setpoint. VAV air handlers also require static pressure control to modify the amount of air delivered. Duct static pressure control is the most common approach, and makes the sixth major control loop integrated with this equipment level controller. A seventh control loop that is becoming common with VAV air handlers is to monitor space pressure and adjust supply air based upon the desire to maintain a positive or negative pressure. With the advent of demand response (DR) and related smart grid sequences, as well as a resurgence of demand limiting, an eighth control loop may be performing a reset-type control sequence on the unit to reduce electricity consumption by changing temperature setpoints. At the same time DR may address duct static pressure, fan speed or a host of other control parameter adjustments to reduce the overall energy consumption by the air handler.

One of the decided benefits to equipment-level intelligent control is the ability to monitor a variety of alarm and safety conditions. Many of these functions are now fully integrated into the BA$^{DDC}$ equipment-level devices as previously discussed. This is particularly true for BA$^{DDCs}$ that are developed and pre-integrated into HVAC equipment at the factory. In the past, these functions were simple control loops that monitor the condition of a contact closure or an analog value. Based on a change in state of the digital point, or an analog point exceeding a set of range limits, an alarm is initiated. As traditional HVAC manufacturers have joined the ranks of market leaders in BA$^{DDC}$, and leaders in HVAC, these sequences and an extensive number of troubleshooting, diagnostic and equipment-related algorithms are fully embedded into controllers. As noted before, access to such sequences would only be available to trained factory technicians using a proprietary workbench software product. Regardless of the deployment mechanism, this functionality is critical to smooth equipment operation. Though these are simple routines, when combined with the temperature and pressure loops above, it becomes clear how complex and diverse the sequences of operations must be that are provided in BA$^{DDC}$ equipment level devices.

Sequence of operations is a complex topic, but before closing there is merit in mentioning the notion of self-learning "fuzzy logic" for BASs. This idea has been discussed for decades, but again there are articles showing up in the trade press about the benefits that could be derived from self-learning BAS. In particular, such sequences could make new smarter energy management strategies possible. For example, a demand response sequence could look at a zone infrared occupancy sensor, and add overhead light dimming to setpoint adjustment during a DR event. Of course there could be tremendous "continuous commissioning" value on a daily basis from self-learning sequences, in this case algorithms that adapt equipment operation based on a series of factors. Consider another example that has to do with telecommuting. Many buildings are lightly loaded on any day of the week because people work from home or are on travel. Enabling sequences can calculate load in the building and adjust heating, cooling and ventilation to how many people are in the building. Some large commercial building operators are already duty cycling elevators when the building occupancy is low. So the value that could be derived for comfort, energy savings, etc. is clear, but the rate at which these enhancements will hit the market is unknown. A recent article by Jim Sinopoli in www.automatedbuilding.com, discussed in the integration chapter, surfaces the idea that the Bluetooth™ signal in smart phones could be integrated into such sequences. The IT integration requirement could add a level of complexity, but this could be the optimum indicator of occupancy. Using an energy and analytics tool, Bluetooth signals could be accessed and analyzed to develop an occupancy trending rule, which could then be used to frame an algorithm that would ramp the building equipment up and down based on actual readings of occupancy based on Bluetooth signals. As with all new technology, the reader should carefully evaluate offerings, speak to references and validate the cost benefit of such offerings carefully before deployment. Given this wide ranging discussion of control sequences including the complexity of ELC control loops, and the potential for self-learning algorithms, the reader is presented with complex technology in the BAS space. This is particularly true when the reader considers that additional sequences may need to be added for control integration.

*Control Integration*

The concept of control integration was briefly discussed under enterprise supervisor (ES) controls and the above vignette on Bluetooth-based equipment control. The same basic definition applies to ELCs as did to ES, except that the focus shifts to appropriate integration at this level. A good example is the concept of temperature reset. One approach to reset is locating a representative sensor somewhere in the space, and making control decisions based upon that single point of information. A more integrated approach would be to monitor space temperature at any number of distributed zone controllers. The BA$^{DDC}$ equipment-level control then might average those values or simply reset to the highest or lowest value. This function is too equipment specific to be delegated to an enterprise supervisor, yet is ideal for BA$^{DDC}$ equipment level control. A little more complex example is morning warm-up. This sequence is used with VAV air handlers that have heating capacity. The air handler will control to the appropriate temperature sensor, and then based upon the control loop will override zone control to force the damper 100% open. This is a truly integrated sequence because control decisions are made at the ELC, but they must be integrated at zone level control.

Control integration at the equipment level is not likely to be limited to interaction with lower level devices. As noted under the enterprise supervisor, these BA$^{DDC}$ ELCs will be integral to carrying building-wide functions such as demand limiting, optimal start/stop and more to come. It has also become much more common for new construction and retrofit specifications to require an enterprise supervisor and other BA$^{DDC}$ devices to seamlessly integrate, at the system level, with BA$^{DDC}$ ELCs, which come from the factory pre-integrated into building HVAC equipment. This form of integration is highly complex and requires a sophisticated integrator. Some believe that BACnet alone enables this type of integration, but is not true. System integration at such a higher level often presents requirements for a host of other software and/or hardware elements. For example, in a phase-2 construction on an existing building there may be new equipment added that has to be integrated with "legacy" equipment from the original constructions. This may require middleware devices that include "drivers" to communicate between the legacy device and a new BACnet device. These concepts will be discussed in more detail under the networking and middleware IT chapters. The primary point here is that integration of BA$^{DDC}$ ELCs has more and more to do with IT, network communication and web services in the 21$^{st}$ century than it does with traditional HVAC control. Yet a thorough understanding of both content areas is necessary for success. Again a thorough understanding of HVAC is critical, but the ideal BA$^{DDC}$ integrator must also be fully versed in IT, web services and data communication disciplines.

That is particularly true for providers' of energy and analytics, dashboards and related technology for buildings.

This is the final topic under BA$^{DDC}$ equipment level controls, and has already been positioned as critical to building-wide control. The functions discussed under the enterprise supervisor for programming, monitoring and dashboard interface necessitate that full internet access is available with the ELC. These devices typically provide complete IP communication capability. Given that statement however, it bears noting that the devices at this level may vary in network and communication capability. These devices often require both integration and dashboard interface capability with enterprise supervisors, and yet it is also common for these devices to be integrated with legacy and other building systems via middleware components. A final topic for discussion here is that the BAS industry has undergone a merger and acquisition frenzy, resulting in major business changes in recent decades. This is relevant to BA$^{DDC}$ ELC devices because, post-M&A, manufacturers have had to decide which BAS product families to make obsolete and which would become the next generation flagship products for that brand. Regardless of specific product decisions made by any manufacturer, companies must also determine how to support the legacy products as well. The relevance to BA$^{DDC}$ ELC devices is that for many manufacturers these devices have become an architecture focal point for integration. It is not uncommon for these devices to incorporate "router" functionality to unify an architecture made up of multiple product families, which as a result of M&A have become one company's post-acquisition offering. The next chapter will discuss these networking concepts in much more detail, but readers should understand clearly whether products installed in their buildings have been affected by such events. If so, readers should explore how manufacturers of those products intend to provide future support and integration.

This overview of current equipment level controls is intended to provide a general view of the technology. Many of these topics will be expanded throughout the book as they relate to big data for energy and buildings. Also of critical importance to building-wide control is the provision of a BA$^{DDC}$ zone level controller.

*Energy and Analytics Impact*

As previously discussed, instrumenting equipment to provide data for analysis has been one of the greatest barriers to entry for energy and analytics. BA$^{DDC}$ ELC devices are where a large portion of the building data exist that can be made available for analytics. Equally exciting is that the enterprise supervisor and the ELCs are most often responsible for virtual points. These virtual points offer a huge source of data for analytics. Virtual point data can be trended and captured over time, so it could include kW peak demand or kWh energy consumption data (whether instantaneous or interval), kW and/or kWh cost data, temperature and humidity data from an external weather station, etc., all of which can provide insight into operating condition data and are excellent data for energy and analytics. It is particularly true that BA$^{DDC}$ ELC devices have access to, and may be trending, extensive amounts of data on conditions, operation, etc. It is important to be aware that the memory for trending data at the ELC is very limited. However, as with the enterprise supervisor, the BA$^{DDC}$ ELC devices should conform to data communications and networking standards, so all of that data can be uploaded to separate onsite storage or to the "cloud." Again data formatting is an issue with various propriety BAS, and other equipment. Ideally systems will migrate to data standards like Haystack. This issue will undoubtedly become more critical as adoption of energy and analytics accelerates and as data standardization efforts mature. There are also individual technology providers who are developing "drivers" and conversion tools to make use of some data. It is important not to oversimplify the idea of "data drivers"; there is a cost for all of this technology, and therefore it will not always be cost effective to access data from all sources. Because of BACnet, and other communication standards, however, it is possible to push vast amounts of data through BAS IP addresses to the cloud or other server.

## BA$^{DDC}$ TECHNOLOGY FOR ZONE LEVEL
## CONTROL APPLICATIONS

As noted there are two types of BA$^{DDC}$ products: BA$^{DDC}$ equipment-level controls which have been discussed and BA$^{DDC}$ zone level control (BA$^{DDC}$ ZLC). Before launching into this discussion of BA$^{DDC}$ ZLC devices, it bears noting that there is actually a third type of product; BA$^{middleware}$, or simply middleware. A full chapter is devoted to middleware, and it has been discussed at some level thus far under the integration topics. As BA$^{DDC}$ systems became more IT oriented and require networking standards and internet access, middleware devices of various types have been introduced to the market and become critical to deployment. The reader is asked to hold that topic for now, as the next two chapters cover the topics of data communications/networking and middleware in more detail. Of course it is perfectly acceptable to jump ahead to those chapters to get questions answered and then return to the BA$^{DDC}$ ZLC discussion as well.

BA$^{DDC}$ ZLC technology has been evolving for decades. As with the BA$^{DDC}$ ELC, these devices may also be part of a BA$^{DDC}$ system offering, or they may be pre-integrated by HVAC/BAS manufacturers. A great recent example of the latter is the variable refrigerant volume (VRV) and variable refrigerant flow (VRF) style heat pumps, which include controls, and have gained rapid market acceptance. That is just one example though, because BA$^{DDC}$ ZLCs have become commonplace with such applications as variable air volume terminal units, or boxes, water source and geothermal heat pumps, and small packaged single zone units. As with the ELC level devices, these controllers are microprocessor-based distributed processing units with full capability for standard network communication.

ELC devices were discussed as both HVAC and general purpose controllers, due to the flexibility of the devices and the applications to which they may be applied. Initially BA$^{DDC}$ ZLCs were primarily focused on unitary HVAC applications. However with the introduction of "wireless networking" and other technology evolutions that will be mentioned, ZLCs are also being used for general purpose control applications. Lighting at the zone level would be the first general purpose application to mention. Lighting controls by ZLC, particularly for dimmable florescent ballasts and LEDs, can be networked wirelessly for scheduling, but can provide space motion detector, infrared or other sensors. As many buildings seek United States Green Building Council (USGBC) Leadership in Energy and Environmental Design (LEED) points, these ZLCs can also be used to deploy control sequences like daylight harvesting to reduce light levels when ambient light is available. As with ELC integration, not all of the BA$^{DDC}$ ZLCs deployed are part of a single manufacturer's product family. Today there's a robust ecosystem of technology companies offering stand-alone ELC and ZLC devices for many applications from demand response to pneumatic thermostat replacement. Some of these applications will be discussed in later chapters, however the focus of this book is on energy and analytics, so the emphasis here is more on the data available from such devices than on an exhaustive treatment of ZLC technology.

BA$^{DDC}$ ZLCs provide distributed processing and BACnet, or other, standard network communication. As a result they require sophisticated control integration sequences to ensure that effective coordination is achieved through the system architecture. To provide more background, these issues will be viewed in similar categories to previous discussions along with pertinent trends that are influencing the market:

- hardware,

- sequence of operations,

- networking and control integration, and

- energy and analytics impact.

*Hardware*

BA$^{DDC}$ ZLC, or more simply ZLC, technology became common due to market demand for control at this level, which ultimately provided economies of scale in production. At the same time microprocessor and associated electronic technology became cost effective, further increasing the desirability of these devices. As with ELCs, these are typically platform devices, which encompass the microprocessor, operating system, non-volatile (read-only memory) and volatile (random-access memory) storage, on-board componentry to sense external conditions and execute control action, software programming and configuration capability and capabilities for data communications networking with other BA$^{DDC}$ devices and, as appropriate, with the internet

Further development of smaller, more powerful microprocessors for ZLC-level devices combined with pervasive deployment of the BACnet standard, resulted in robust development of these products. ZLCs are typically tailored to a specific application and fulfill the definition of direct digital control by enabling control intelligence located at each piece of controlled equipment.

The nature of ZLC products imposes requirements on these controllers that are applied to zone or unitary equipment. As with ELCs, input/output capability must be provided. BA$^{DDC}$ ZLCs have less input and output (I/O) capability than ELCs and offer more limited expansion capability. Yet they must be well suited to a variety of applications, such as variable air volume air handlers and water source heat pumps. At the same time these devices also must be cost effective for those applications. In addition, there are several hardware related areas that are equally important. Among these are component packaging, physical dimensions, and ability to operate at outdoor ambient temperatures.

Input/output (I/O) requirements are generally less stringent than with the ELC due to the limited number of points required with zone applications. Overall there will typically be fewer points, however analog and digital inputs and outputs are still necessary. This is because most zone applications will require both monitoring and equipment control. Control is accomplished through digital and analog or pulsed modulating outputs for such equipment as dampers and valves. Again manufacturers typically develop one ZLC hardware platform and use it with all zone applications. This is accomplished by changing software in the controller. Software approaches related to this activity will be discussed further under sequence of operations.

With specific regard to inputs and outputs (I/O), an important zone hardware design issue is the available number and type of I/O channels or points. Designers must first determine the number of points that are appropriate for monitoring and control. Consideration must also be given to whether a fixed number of I/O will be available, or if expansion is to be provided. Expansion I/O may be critical if multiple applications are to be accommodated by one platform. Expansion is usually provided through boards that may be added to the base device or via a module that is applied externally to the base device. There will be trade-offs associated with any I/O expansion scheme. Two key points regarding I/O are that ZLC sequence programming capability may be limited because the device is generally intended for one relatively simple piece of equipment. As integrators consider adding more I/O they should evaluate device capacity to support code necessary for controlling those points. At the same time, with the growing application of general purpose control, programming capacity can be an issue. There is logic in using one ZLC for a particular zone and controlling HVAC, lighting and possibly other ancillary loads. There will be more discussion of these concepts under sequence of operations.

ZLC hardware design issues that are of particular importance involve the requirement for a full line of products including the controller, sensor or thermostat, static pressure measurement device, damper or valve motors and other ancillary gear depending on the application. These devices must be optional and fully compatible with each other, and the ZLC, regardless of application software. As with the ELC, the expectation is for full flexibility in selection of these components. Given the growing application of ZLC general purpose control, this is even more important. Even with target applications like VAV, the diversity in unitary equipment requires wide ranging options for electronic sensors, and as unlikely as it may seem there still may be a need in some cases for standard interface to pneumatic output devices. Across the board, ZLC devices and ancillary components are highly price sensitive. The primary recommendation here, as with BAS in general, is that the reader should be careful not to select vendors based upon price or brand name alone, but to evaluate service as well as product cost.

Component packaging is critical because zone control products tend to be self-contained in nature. Traditionally ZLCs tended to be applied by original equipment manufacturers (OEMs) in the factory, and each manufacturer offered a variety of control products. There are actually two distinct issues that are involved: packaging as it relates to easy application of zone products and packaging as it relates to physical design. As has been discussed there is a robust marketplace of third-party BAS device and component providers who offer solutions to the market. The key is for readers to educate themselves and ask the right questions during selection. VRF and VRV are a good example of this point. These heat pumps are deploying more than high efficiency solutions, they are deploying new approaches to system design. To achieve higher efficiency, direct current (DC) motors are used for fans and compressors to allow a wide range of modulation and tailor equipment loading to space comfort requirements. This makes the units very efficient but it also requires a controller that can send DC control signals to device components.

Application of ZLCs has become pervasive for all types of unitary or zone HVAC equipment including VAV boxes, air, water or ground source and VRV/F heat pumps, self-contained single zone packaged roof top units, packaged terminal air conditioners (PTACs), fan coil and most other units of this type. Using VAV as a component packaging example, it is now standard to provide one easily installed package containing the controller, sensor or thermostat, duct static pressure measurement device, and output devices for dampers and valves. Original equipment manufacturers (OEMs) of zone HVAC demand simple control interaction. More to the point, these OEMs have achieved significant market share in recent decades as leaders in the manufacturing of both the HVAC units and the controls or BAS. As will be discussed under integration, full market acceptance of BACnet coupled with significant advancement in system integration has resulted in a new BAS paradigm. There is a wide ranging belief among consultants and building owners that pre-integrated OEM ZLCs can "interoperate" with BACnet ELC and enterprise supervisor devices manufactured by other OEMs. In fact, this is possible. However, all OEM devices and all system integrators are not created equal. Achieving BAS interoperability requires a very sophisticated system integrator. Perhaps the most important point is that the ideal design process will include a system integrator, and procurement selection must evaluate the integrator as carefully as the price. Low price is not always, or even often, best.

The physical design aspect of ZLC packaging relates to the concept of ensuring that a control meets the space constraints of the application. This is actually a hardware design and selection issue. In fact, space constraints are always a concern with any control product, but are more important with the ZLC because very limited space is available. VAV boxes, for example will vary greatly in size based upon the zone CFM requirement and the type of box. A pressure dependent box with no fan or zone reheat and low CFM requirements would likely afford very little space for the ZLC. It is more common today to simply mount these devices on the outside of the box, but the designer should confirm that is an option. Also, both OEMs and building owners will require that the same ZLC be applied for all size boxes, though each for different reasons. The OEM is concerned with standardizing all facets of production including installation, manufacturing procedures and test/checkout for the ZLC. In this case the building owner is concerned with interoperability and system control integration for all of the VAV boxes and for the entire BAS. Further, as discussed above, VAV boxes are not the only application that will be controlled by a ZLC. As a result, use of any ZLC will be dictated to some degree by its ability to meet space requirements for any intended application.

Ambient temperature (where the ZLC is installed) is another important hardware consideration. Again this is a critical hardware design and selection issue, as is physical dimension. Key to the discussion of ZLC products is that the variety of applications for these devices requires that they meet a range of operating conditions. In nearly all cases, ZLCs which are mounted with VAV boxes need only operate under standard interior ambient temperature conditions; but, single zone packaged units, which may be fitted with the same ZLC hardware platform, are likely to be exposed to exterior ambient temperature and humidity conditions. Another common ZLC general purpose application is exterior LED lighting, which may also require outside placement. As may be obvious, this dramatically increases the requirements, and perhaps the cost, for these devices due to the potential for application in Alaska or Florida. Also the control business is global, so ambient conditions must be considered worldwide; therefore, a ZLC could conceivably be applied in arctic or equatorial zones.

*Sequence of Operations*

As discussed under ELCs, the approach and style for programming control sequences varies dramatically by manufacturer and type of equipment. This is true of any HVAC or load control application, yet there are added distinctions between the ELC and ZLC. At first look, a reasonable case could be made that ELCs must maintain more complex and sophisticated control sequences than would be encountered in zone applications. In many cases this may be true, yet the zone device must be able to meet the requirements of all applicable equipment. Throughout this chapter the VAV box has been the example; consider again this piece of equipment. There are a number of VAV implementations, with pressure dependent being simplest and others such as pressure independent dual duct applications being more complex. As an extreme case the pressure independent system will be considered, but for simplicity's sake we will focus on single duct. Note that the sequences outlined are only an approximation and used for example rather than as a direct control application.

Single duct pressure independent systems also vary in implementation. For example they may include fan-assisted systems, and may also provide reheat. Consider one of the more common systems, a single duct application with fan assist and one stage of electric reheat. Assume that control is provided by a dedicated ZLC, also typical. This system would require two primary control loops: space temperature control and static pressure control. The space temperature control loop modulates a damper to vary the amount of air delivered to the space and satisfy a control setpoint. A call for heating in the space would also sequence the electric reheat. The same algorithm would also provide occupied and unoccupied setpoints along with a sequence for the unoccupied mode. The unoccupied mode sequence would very likely provide at least one set of control parameters for the fan and reheat. It must also allow for starting the VAV air handler to satisfy cooling requirements. Operating concurrently with the temperature control is a static pressure loop. This loop is intended to allow for variations in cubic feet per minute (CFM) delivered to the box. Pressure control ensures that the space does not suffer from noise or discomfort due to variations in delivered CFM from the air handler. This sequence controls based upon data from a duct static pressure measuring device, and will integrate response to CFM variations with the zone damper control sequence.

The reason for this somewhat detailed discussion of one application is to raise the reader's awareness of zone control sequence complexity. Given that zone applications typically involve smaller less sophisticated equipment, and smaller numbers of I/O, it would be easy to assume that sequences are simplistic as well. Rather, sequences tend to be complex, and there are also interactive sequences to be considered. The same VAV box control requires networking communication to allow integration with the air handler (AHU) for simple functions like unoccupied override to start the AHU. Slightly more complex strategies might include reset of discharge air from zone temperature and building pressurization control with box and air handler integration. This means that it is entirely possible to use the full power of a microprocessor in the ZLC to provide sequence and communication/networking/integration requirements, which will be discussed further below.

**Networking and Control Integration**

BA$^{DDC}$ ZLCs must be networked to execute the control sequences that are carried out between modules at the zone level, as well as for integration with ELCs and the enterprise supervisor. These concepts of networking and integration are covered under the same heading for ZLCs because these devices are highly reliant on network interaction for effective building operations. At the same time, the control sequences at the ELC and building-wide rely on data from the ZLC level to provide effective operation. Networking and integration are therefore completely reliant upon one another. For example, ZLC control integration is essential because it is not possible to terminate every necessary point to this panel. At the same time, manufacturers vary on whether they choose to provide every control sequence at the ZLC. Consider scheduling for time of day on/off or setpoint functions, some ZLCs have clock functionality in the controller, while others do not. Actual time of day (TOD), or TOD and the ZLC schedule, may be maintained elsewhere, and a network broadcast from the ELC or enterprise supervisor sends this information to the controller. Typically this is done with a data point indicating to the ZLC that it is now in either occupied or unoccupied mode. These and other data points fall into the "system data" or "virtual point" category. However, it may also be necessary to implement building-wide control such as optimal start or demand limiting by issuing virtual points that initiate sequences through the ZLC.

**Energy and Analytics Impact**

This topic has been covered under enterprise supervisor and BA$^{D-DC}$ ELC and is equally appropriate for the ZLC. Given the sheer volume of condition and status data available from tens, hundreds or even thousands of ZLCs in a building, it is extremely important to make this information available to energy and analytics tools. These data are crucial for continuous commissioning and enterprise energy management because data from the zone level is the best indication of how well a building is being operated. Operations in this case could focus on maintaining space environmental conditions that meet the mission of the building, but equally important could relate to operations and maintenance for equipment life as well as energy efficiency to optimize utility costs. Overall ZLC data are essential to analysis of these issues, as well as indoor air quality, measurement and verification for energy efficiency and much more.

As previously discussed, instrumenting equipment to provide data for analysis has been a major barrier to entry for energy and analytics technology, and zone equipment presented one of the greatest challenges in this area. Combine the numbers of devices mentioned above with the challenges of cabling for all of the instrumentation. These challenges were equally limiting in getting automated control to the zone level as well. This obstacle to gathering analysis data from the zone level is gradually waning, because ZLCs are being applied almost universally for HVAC. Also new ZLC features have been introduced including general purpose and lighting control. Whether these devices are networked over a hardwired or wireless network, their data can be communicated to a service, or cloud, via an enterprise supervisor, even if a middleware device is required. This is particularly important for ZLCs, because memory for trending data is extremely limited. Standard communication networks and data formatting remain a very important topic as previously discussed.

## ACTIVE ENERGY AND ANALYTICS

This topic has been discussed elsewhere in this book and will not be covered in detail here except for one important point. With a more detailed understanding of BAS technology it should now be easier for the reader to understand the complexity associated with acquiring data for energy and analytics. Equally important, the reader should understand the complexity associated with using the data from the analysis performed by this technology, to optimize building operations. In many cases new sequences may be developed to further automate this optimization, but it must be an active process. The operator must:

- Actively perform analytics,

- Actively develop strategies and sequences to improve overall operations based upon the results of the analytics, and finally

- Actively perform continuous improvement to ensure that the sequences achieve the intended result.

# Chapter 11

# Introduction to Digital Communication for Building Automation and DDC

The technology covered in this book is wide ranging by intent, because of the extensive and diverse set of "systems" that are expected to be providing data input for energy and analytics tools. These systems can range from building automation and related technology, as described in the previous chapter, to energy meters, programmable logic systems for process applications, point of sale and a multitude of other on-site computer based technologies such as Bluetooth™, but they also include the myriad data sources from the internet, cloud, etc. The diversity of these on-site systems is complicated by the fact that, in many cases, they were designed to operate independently. Therefore networking and data communications between them, and for data access from them, requires an understanding of the technology covered in this chapter. It is highly beneficial to provide a rudimentary understanding of data interchange technology as a resource for future use. Also of interest, is that the technology and many of the data communication standards discussed here are integral to the internet and data interchange between all computer systems and the cloud.

The next chapter will focus more on middleware, the internet and related web-based technology. This one addresses the underlying data communications and networking technologies and architecture. These networks are proprietary (developed by a company and kept as private intellectual property) or based upon standards (published and open for used by any company such as BACnet™). The data communication technologies that underlie energy, buildings, process, campus, plant and other systems are very important to developing strategies for accessing the data from those systems initially, as well as to ensuring the long-term reliability of those data streams.

Many readers may be unaware of how precarious system inter-actions can be, and how important it is to establish rigorous oversight of the data communication channels to ensure "data resilience." Cyber security is a major issue as well, and system operators must remain vigilant to avoid threats to building networks from data sharing via the internet. Developing systems that deploy cyber security strategies while data are being captured from decentralized and integrated net-works is critical to managing risk. As the next two chapters unfold, it will become clear that operators must begin with a robust information architecture design. They must then develop data access systems that apply the same level of analytics to those architectures, as to the ap-plications, to ensure that a continuing and reliable stream of data is always available. The author refers to this as data resilience.

Data resilience is a concept that becomes even more critical in light of the continuing trend for these systems to utilize web services and dashboards for all operator interface. Information interchange for many specialty systems is migrating to the internet, or web, so it is crit-ical for customers to address cyber security and data resilience for these systems. Despite the fact that a network of businesses and independent bodies is overseeing internet security, the recommended efforts focus on specific uses of the internet and specific web applications or ser-vices. As a result, the task of maintaining the viability of data streams falls to the operator's side of the web interface. This is critical because most of the underlying technology still relies on systems that use a host of legacy and standard data communication protocols that must access the internet through some level of device that is capable of being assigned an internet protocol (IP) address, and deploying some level of web services. As has been discussed throughout this book, access to data from many of these energy and building systems requires interfac-es, gateways, drivers or other mechanisms to push that data through an IP addressable device to the cloud, etc. This means that foundation-lev-el knowledge in data communications and networking can be of great value to the reader, and that is the purpose of this chapter.

## INTRODUCTION TO DIGITAL COMMUNICATIONS

This chapter will cover a wide range of topics designed to provide the reader with an understanding of data communications technology in general, including system architecture. At times these technologies are referred to as local area networks (LAN). They may also be called "sub-nets" or "sub-LANs" because the overall communications architecture in use here entails multiple levels of networking. At this sub-LAN level, the focus is on data communications for building automation and other building systems such as security, computerized maintenance management systems, energy metering or enterprise energy management systems. Multiple topics will be covered to address the protocol and networking standards related to these systems and to energy and analytics overall. Initially it is important to start with a discussion of protocol as it relates to data communication networks. For all the reasons stated thus far, this should be useful background for the reader as they consider the impact of this topic on energy and analytics. As a result the reader may need to research elsewhere for further information about data communication and related information on the local area network industry.

The basis for data communication and network is protocol. A communications protocol is a system of digital rules for data exchange within or between computers, and those rules also establish the architecture of a network. As such, protocol is a somewhat simple concept, yet the interrelationship between systems and the requirements for energy and analytics make it a large and complex topic. The focus here is on relating existing data communication and BAS technology to analytics. The author believes that an understanding of these topics will assist the reader in making better everyday decisions about operations and system design. In the end, the goal here is to exchange data in the form of computer messages, and the rules for that exchange are established by the protocol. Since the exchange of messages that contain meaningful data is the goal, it is helpful to understand the concept of messaging to gain an awareness of methods used to exchange messages within computer-based systems. The discussion will then progress through a series of protocol-related topics, each of which has a bearing on data access and data resiliency.

The term protocol will be used extensively with a focus on networks, because it is the author's contention that system networking is at the heart of energy and analytics. Networking provides the basis for BAS devices, and other computer-based systems, to communicate with one another and to communicate data to the outside world of analytics systems, etc. through the internet. Network protocol standards for BAS, energy and other building/facility systems are central to providing on-going data access necessary for operational analysis, fault diagnostics and detection (FDD) and continuous commissioning. To address these technologies, this chapter will define and discuss protocol, starting with data messaging. After a discussion of how messages are constructed, which would be necessary to request or access data, the chapter will follow with a complete review of the data communication, protocol and networking technology that exist to enable this data messaging process.

The chapter will define and describe key concepts, with expanded discussion of protocol and networking technology as it relates to energy and building data for analytics. This chapter will also cover the topic of system architecture, which was discussed with regard to BAS in Chapter 10. Perhaps of greatest import to the evolution of energy and analytics topics however, is to be aware that protocol and networking technology are mature in the IT space, as well as with buildings where standards like BACnet™ have been deployed for nearly two decades. These standards may require adaptation and evolution from time to time, but they offer a robust foundation for application to creating energy and analytics tools. Ultimately however, the informed operator will recognize that some degree of system integration will likely be required, so it is important to have a rudimentary knowledge of protocols and protocol reference models. The differences between protocols and protocol reference models will become clear as we discuss these topics further. At its most basic level however, data are the key requirement, and data must be communicated between computer-based systems and the cloud through a message.

DATA MESSAGES

The concept of messaging is essential to one of the key demands in our industry, the need to have access to the information or data that are available from many different systems. Those data are structured in the form of messages with formats that are dictated by a protocol. All the critical control and operational trend information residing in various energy and building systems are data, and at their most elemental level data are made up of binary digits.

These binary digits, often called bits, make up data consisting of two elements "0" and "1." These bits are assembled into a message with two basic types of data: control and user information (data). Control, not to be confused the BAS control application, conforms to requirements of the protocol and ensures that the information, or data, arrives at the proper end location. These control data are essential to managing individual transmissions, as well as interaction between the various computer-based devices (called "nodes") on the network. The user information, which is the primary reason for networking, may contain a space temperature, an energy consumption or demand value or any number of other types of information. The concepts associated with converting the message to an electrical signal and transmitting it over a "medium," or wire, are also the subject of published "standards" but are not germane to the focus of this book. The primary focus in this chapter is the information (data), the form they take and the process necessary to communicate them between nodes.

It may be evident at this point that the key to sharing information between two different computer-based systems, or between those systems and a cloud-based system, is ensuring that the message formats are identical. There are a number of other considerations as well, because the message format is dictated by other factors including the type of transmission that is employed. As noted, the focus here is on the data themselves and the process employed through a communication protocol to transmit them to another node on a network, another system, or to the cloud via the internet. The approach taken will be to describe message format as a concept and then to discuss the protocol that transmits that format. As an example of message formats consider the synchronous transmission process, shown below, as depicted by Uyless Black in his book *Data Communications and Distributed Networks**. The author has included Mr. Black's explanations below and incorporated, where appropriate, any issues which must be considered for messaging for analytics.

**SYN Characters**

These are timing characters used to synchronize the timing of communication between two nodes residing on the network. These characters also act as start and stop bits to identify the beginning and

---

*Uyless D. Black, Data Communications and Distributed Networks, Second Edition, Prentice Hall, Inc., Englewood Cliffs, N.J., 1987, p. 252

end of a message. For analytics, such characters would serve the same purpose, yet it is important to note that microprocessor based devices on the network will likely not initiate communication. In most cases data will be "pulled" from network devices on a higher level analytics system device, therefore it is most important for that device and the analytics engine to execute the same functions.

**Header or Address Field**

It is critical to identify the transmitting and receiving nodes on the network. This is done through a series of characters, including numbers or letters, for each controller. In most cases there is a requirement for the node address or name to be unique for each controller on a network, and this rule holds true for energy and building networks. This address is normally for network use only, and a number of techniques are employed. Probably the most common method, however, is through a series of software or dipswitch settings that correspond to a software setting or address that is held in device memory. A corresponding address must be programmed in a device such as the enterprise supervisor discussed in the previous chapter that will be used for commissioning or monitoring the node. Allowing users to identify controller nodes by a definable name is also important for communication over the internet through an enterprise supervisor type of device.

**Control Characters**

Again it is important not to confuse these characters with the process of controlling equipment that is carried out by a BAS. The job of these characters is to maintain the flow of data, and ensure that it is in the correct sequence. These characters are essential to communication, and will be determined to a great degree by the protocol employed with the control network. Control characters employed and their placement will vary dependent upon the protocol and its implementation; however, in some cases the user data will be preceded and followed by control characters. Many of the efforts underway to allow connectivity must address the control aspects of messaging to allow multi-vendor networks to exist. There will be a great deal more discussion of the requirement for control of messages under the protocol section of this chapter.

**User Data**

In the information technology (IT) industry, these data consist of one or many fields, created by an operator or by the output of a computer program. IT industry user data are associated with the computer application. In the energy and analytics industry, user data can encompass a wide range of content, and they are also related to the application. Yet control user data may be even more critical than with IT due to the real time control decisions that are made on a second-by-second basis. At this book's writing, the intent of most energy and analytics technology is to evaluate a wide range of performance data about buildings, to analyze those data and to provide results. Those results are often reports on building performance, but they also often include recommendations to improve performance. Energy and analytics help operators to make decisions to effect a change and improve performance, but they typically do not execute any type of automatic action. In summary, data are "accessed" and analyzed by the tool to determine whether a recommendation can be made to improve a process. Consider a fault diagnostics and detection (FD&D) application with the accessed data approach, where the data accessed and the analysis conducted indicated that the mixed air function of an air handler was not operating properly. The recommendation to correct that issue might be to take any number of steps including physical activity such as replacing an actuator, programming activity such as modifying sequences, etc.

"Accessed data" from a variety of systems are critical to energy and analytic tools. However it is conceivable that systems will be developed, particularly for energy, which might also use a second type of control user data, "control integrated data." As discussed above, accessed data are generated by systems like a BAS, and may be access through the analytics systems requesting or "pulling" data, i.e. temperature information, control status, etc. for analysis. The purpose would be to ensure proper operation of the air handler by conducting FD&D on the equipment. Most often those data will be part of a larger data interchange between a site and a cloud-based analytics tool, but it is conceivable that a local operator could place the request as well. The idea behind control integrated data would be to update the information on a real-time basis, conduct the analytics and, if appropriate, to integrate a local site response. Given the fact that smart systems are installed in buildings, which can respond to an electronic message, this is achievable. As noted above, energy would be an ideal application and these types of strategies are already being used for demand response with electricity.

**Error Check Data**

Another essential aspect of messaging systems is to utilize schemes for error, though these will not be described here. The key issue is that data must be checked to ensure that the correct information is received. In this way, a high degree of accuracy may be verified with both accessed data and control integrated data.

Given this discussion of messages the reader should have a better understanding of the some of the concerns associated with ensuring that messages from disparate systems and other sources are accurately received by energy and analytics tools. A number of other concerns must be addressed to transmit this information between machines and allow connectivity. Another term commonly used rather than connectivity, is interoperability. Interoperability is appropriate in this case because the communication may be two-way. The goal of interoperability for energy and analytics is data access, and the definition will be that communications between the devices is successful and that data are accurately transferred between devices. When integrating BAS systems there are additional requirements, but in this case interoperability is the requirement, and it will be the focus during this chapter's discussion of protocol.

**Introduction to Data Communications**
**and Networking Standards**

The discussion of networks and protocols throughout this book will be limited in scope. Yet it should be evident by this point that the issues of networking and protocol are almost inseparable. It is also true that standard protocols, often simply referred to as "standards," must accommodate remote communication between the internet, a variety of BAS, metering and other hardware. Equally important, a proliferation of web-based devices are being applied for energy, buildings, etc., creating a network architecture of systems and individual devices or controllers. As a result the protocols, or standards, must accommodate both controller-to-controller communication and remote network interface via the internet. The concept of using the protocol to provide a means for integrating communications and data access between different devices, manufacturers, etc. also assumes a network of some type. There must be a means for establishing communication between multiple microprocessor based controllers, and ensuring that guidelines are established for the exchange of information or data. The means for sharing that information is a network, and the rules which define the makeup of the data are the basis for a protocol.

This chapter is intended to expose the reader to basic protocol and networking concepts. More extensive discussion of these concepts is available in texts dedicated to this discipline. The goal here however is to provide an overview of the technology with key buzzwords and acronyms, so that readers are able to analyze the requirements and complexity associated with data access via networking. Understanding networking requires an understanding of some very specialized terminology and the ability to evaluate multiple different standard implementations. Networking is also the basic building block for the internet, and is therefore a vital aspect of any modern computer or information technology (IT) system, including specialized systems like BAS. There has been a tremendous amount of effort applied to networking between and among systems, but this discussion will be confined to terms that have relevance to energy and analytics, or are relevant to its requirements.

Given that prologue, this chapter will make mention of general IT industry terminology, and will quote briefly from a *PC Magazine* article written by Frank Derfler, Jr.(1). The article may be dated, but the terms, as outlined by Derfler, are defined in exactly the same manner today as then, and are the same for the energy and buildings industry as for IT. This chapter provides definitions of communication terms as they apply to both IT and energy and analytics. In Mr. Derfler's words in that article, "before you can understand networking, you've got to speak the language." (Derfler, Frank J., "Networking Acronyms and Buzzwords," NY, NY, June 14, 1988.)

Derfler also notes that "you need a structure to hang the acronyms and buzzwords on, "therefore, you first have to know about the ISO and its OSI model. The International Standards Organization (ISO), based in Paris, develops standards for international and national data communications. The US representative to the ISO is the American National Standards Institute, or ANSI. In the early 1970s, the ISO developed a standard model of a data communications system and called it the Open System Interconnections (OSI) model. The OSI model, consisting of seven layers, describes what happens when a terminal talks to a computer or one computer talks to another. This model was designed to facilitate creating a system in which equipment from different vendors can communicate."

These data communications and IT industry notes are key to this text because the BAS industry, as a major source of analytics data, has focused a great deal of effort on the OSI model. The standardization process in buildings is led by the American Society of Heating, Refrigeration and Air Conditioning Engineers (ASHRAE) with its Communication Protocol Standard Committee 135P. That committee has elected to use the OSI model as the basis for a standard protocol.

## PROTOCOLS

As discussed thus far, whether it's wired or wireless, most data communications today happens by way of packets of information travelling over one or more networks. Before these networks can work together, though, they must use a common protocol, or a set of rules for transmitting and receiving these packets of data. Many protocols have been developed. One of the most widely used is the transmission control protocol/internet protocol (TCP/IP), often associated with the internet. TCP/IP will be discussed in more detail in the next chapter. The open system interconnection (OSI) model however is a more generic protocol model that is very helpful in describing network communications, and is also useful for comparing and contrasting different protocols. Most of the buzzwords discussed here will be protocols. Frank Derfler likens protocols to "the signals that a baseball catcher and pitcher exchange for pitches, protocols represent an agreement among different parts of the network on how data are to be transferred. Though you aren't supposed to see them and only a few people understand them, their effect on system performance can be spectacular. A bad protocol can slow data transfer, but a good one can make communications possible between dissimilar systems."

A protocol consists of critical elements which define the format, structure and timing of data transmission. Format is essential to defining the organization of the data that are sent. Data structures are needed for coordination between devices and for handling data. Protocol timing involves matching transmission speeds to allow communication between devices with varying capability (for example the reader has likely experienced a time when number of users on a WiFi network slowed refresh speed). Timing is also implemented within the protocol to ensure that data are transmitted in the proper sequence to ensure that they are meaningful when they arrive at a receiving station.

The open systems interconnection (OSI) model is a means for standardizing the process of developing protocols. The reader should note that OSI is not a protocol, but a model to follow in writing new protocols. This model is intended to allow standardization by segmenting the tasks that a protocol must carry out into functional areas, called layers. The OSI model contains seven layers as noted in Figure 11-1, and has been used in the development of many protocols and protocol standards throughout the world. This model has been the choice of many organizations within the data communication and IT industries. Having been in existence since 1984, it has also been proven by experts in that industry.

The rationale above, along with a number of other factors, was taken into consideration by BAS industry experts in selecting a method for developing a standard protocol. The OSI model and associated terminology are a focus for this chapter, with the intent to provide a rudimentary understanding of networking and protocols.

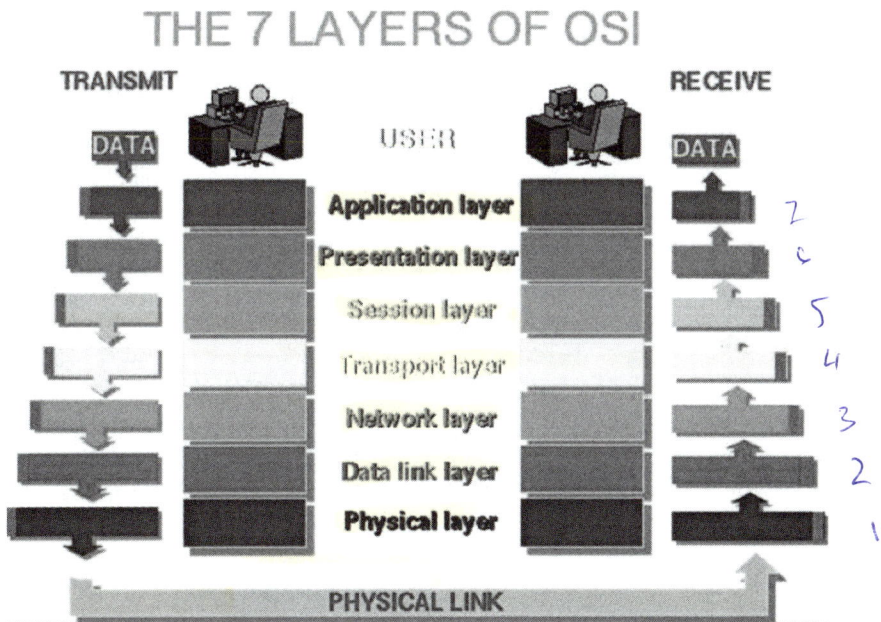

Figure 11-1. OSI Reference Model Layers

**Physical Layer**

With the exception of this layer the OSI model consists of modular blocks of computer software. The physical layer is the only hardware element of the model and it is characterized by electrical connections, wiring and signaling, often termed "media." All of the higher layers must communicate through the physical layer. Examples of physical layer media are: Twisted-pair wiring, RS-232C cable, fiber-optic strands, and coaxial cable.

An interesting aspect of the OSI model is that each of the layers which comprise the overall protocol may be a discreet area of expertise. An example of such a standard that has been established within the physical layer is RS-232C. RS-232C is a wiring and signaling standard that is implemented via cable connections between machines or devices. These cables are connected by pins, and the standard defines which pin does what, and when a voltage level on a wire represents a 1 or a 0.

The physical layer is critical to network communication, and carries electrical signals that represent data from the higher OSI layers. Without this layer there can be no communication. Yet the upper layers are absolutely essential if there are to be meaningful communications, i.e. data access.

**Data Link Layer**

Given that the electrical media are in place, software must be implemented to control the information passing between network members or those members and an internet protocol (IP) connection or address. The OSI model segments this task into the data link layer which assembles characters into messages, controls the transmission of those messages over the media and then checks them before sending them on to the physical media. This layer sometimes transmits messages to notify the sending node that the data have been received. But a "receive message" is typically not sent until some type of error checking has been done to verify that the data are correct. This is a very important topic for the energy and analytics industry because critical data cannot be lost. Fortunately however, most devices have enough buffer memory to be able to resend, providing these message confirmations are timely. As with the physical layer, the data link layer also has standards or protocols that are implemented. Among these is the high-level data link control (HDLC).

## Network Layer

This layer is necessary with large networks which may allow many options for transporting blocks of data from one location, typically a workstation or server in an IT local or wide area network, to another. Note that this is a good example of the interdependency that OSI incorporates between layers. These blocks of data are assembled by the data-link-layer, and the network layer allows network connection between nodes and routes the data transmission through one of a variety of paths. The actual physical pathway determined for the data will be based on network conditions such as nodes that have failed, priorities of service, and other factors. Again, this layer's function is tailored to larger IT local area networks (LANs) in IT applications.

## Transport Layer

This layer performs a similar function to the network layer, except it is focused on node-to-node interaction within a smaller system, such as a LAN. The key functions of transport are to enhance data transmission reliability. This is done by ensuring that the information is passed between specified nodes, and if a failure occurs, determining another node that can pass the data. A key feature of the transport layer is that it can save data blocks if there is no appropriate transmission route on the network. This is very important in the case of a failure in the communication network, or a necessary node, to communication. The transport layer also implements data integrity features through sequence that check the data. At this layer it is possible to dissect blocks of data and confirm that information has not been corrupted.

## Session Layer

This layer establishes the requirements for two network members to carry on a communication session. The session layer defines the rules for interaction between these two nodes, and perhaps the most common example of a session is between two workstations or network members on a computer network.

The primary intent of this layer is to provide a way for entities in two separate devices, potentially employing distinct protocols, to organize and synchronize their dialogue and manage the exchange of data. This is particularly important if those two entities, perhaps PCs, are operating the same application software. For example it may be necessary for the server to provide data handling for user files, while portions of the application software may reside in a workstation. Among the key functions provided at this layer, for communication over the network, are name recognition, logging, administration, and other similar functions.

### Presentation Layer

This layer, as the name might imply, is oriented to the display of information. This information may be in the form of graphic- or text-based data, and still it is in the area of the presentation layer. Data passed from the lower layers must be formatted for presentation by the network software. This formatting may involve graphic representation or may require that data from another protocol be converted for presentation in a given system. A variety of related functions such as printer interface and formatting for special displays, etc. would also be done at this layer.

### Application Layer

This layer is perhaps the most critical for discussion with any network, including BAS. The application layer tailors the process of data communication to specific user applications. It provides an interaction with the network software to carry out the functions necessary to achieve the network's purpose. This direct interaction with the network software is essential for interaction within any network, and for interface with the outside world via the web. The network software defines media access by the network members and provides any custom functionality that must be accomplished.

It does bear noting within this discussion that the application layer also duplicates, at a more specific level, functions that are provided at lower levels. These functions include network access, flow control and error recovery. For example, at this level, network access may be oriented toward a user password and identification code from a dashboard interface. This code will be defined in detail within the dashboard with regard to the number and type of characters. Based on an authorized code, the application layer might limit or disallow communication with the network. In much the same way, more specialized methods of flow control or error recovery might be implemented at this layer.

As Derfler states, "That's it—the top of ISO's OSI model. The concepts are pretty easy, but dozens of committees are working to define standards for little pieces of each layer, and great political fights are being waged over whose ideas should prevail." The discussion of protocols is not complete without a discussion of network software. There are two key components of this software, as noted above, media access control (MAC) and application software. The source of MAC standards is the Institute of Electrical and Electronics Engineers (IEEE).

## MEDIA ACCESS CONTROL (MAC)

Derfler also states that, "the Institute of Electrical and Electronics Engineers (IEEE) developed a set of standards describing the cabling, electrical topology, physical topology, and access scheme of network products. The committee structure of the IEEE is numbered like the Dewey decimal system. The general committee working on these standards is 802. Various subcommittees, designated by decimal numbers, have different versions of the standards." Therefore the standards that have come out of this effort are often referred to as 802.X.

This chapter will provide further discussion of MAC standards under the architecture heading, but a brief overview of important topics is covered here to emphasize the significance of these standards. The reader should note that these standards directly correlate with the 1st and 2nd OSI layers. The concept of MAC is to standardize on the process of allowing a network member to gain access to the communication path. For example, there are two common approaches taken with BAS systems, and these have been defined to date by the sophistication of the control product. Media access is critical to BAS because it determines to a great degree the reliability of data communication. Control system information, particularly those data which are needed for analytics, is highly critical and losing those data would significantly diminish the value of reports and recommendations. Access control protocols that have been used in this industry are of two types: Carrier sense multiple access with collision detection (CSMA/CD) which has been used in more sophisticated systems, and token passing which has been implemented to a great degree with BAS products across the board. These MAC standards have been commonly implemented for proprietary control networks.

CSMA/CD is covered by IEEE standard 802.3. The approach to media access has been well established and proven in the computer industry. In his book on local area networks, Thomas W. Madron states that 802 standard CSMA/CD systems, particularly Ethernet have been the most widely deployed and supported networks in the computer industry*. This approach is particularly well suited to high-speed networking with requirements for nodes to have rapid access to the media. As Madron and others have discussed, the CSMA/CD concept has often been compared to a group of people talking. In such a situation each member of the group listens, and waits for an opportunity to participate in the conversation. This is generally effective, yet there are occasions when two members of the group will attempt to speak at the same time. This results in the group receiving a garbled message. In much the same way, CSMA/CD network members listen for the media to be idle, and then transmit. This allows access to the path with a minimum of delay, yet a procedure must be establish for reinitiating communication when two nodes attempt to communicate simultaneously. The process often called contention, requires that the network members listen for a data collision, notify other members that a collision has occurred and then retransmit in order.

The second means of access control to be discussed, and perhaps the most common to date in the controls industry is token passing. Token passing is covered by two IEEE standards, however the most appropriate to BAS industry is 802.5, the Token Bus. Access control in this case can most commonly by compared to a relay race. The token itself is a piece of information that gives one node at a time authority to communicate. Much like the baton in a relay race, only one runner at a time has the field for a given race team. Token passing is often referred to as a collision avoidance technique because it disallows the possibility of simultaneous communication. As noted, data collisions are not acceptable for BAS or any other devices that rely upon input data to analytics systems, and this has been one of the factors in broader implementation of this approach.

NETWORK SOFTWARE

In conjunction with the protocol and the media access control standards, software must be implemented to provide network functionality. In IT networks it is essential to implement complex software

---

*Thomas W. Madron, *Local Area Networks, The Second Generation*, John Wiley & Sons, NY, NY 1988

packages to interact with the application layer of the protocol, and accomplish desired user functions. These functions include accounting, database management, access and security control, as well as the network operation of a variety of application software packages. In BAS networks, a similar requirement exists, though most of these functions will be carried out at the controller or enterprise supervisor. The enterprise supervisor is a central communication master that manages communication and data interaction between panels. An enterprise supervisor may also implement data sharing and other functions to be discussed in the next section. Most important to this discussion is that this software must be transparent to the user in control applications, yet it must not be taken for granted as an integral part of the network system.

## SIGNAL BOOSTING AND INTER-PROTOCOL NETWORKS

A function of communication networks that are hardwired, is that signals can travel only limited distances over the media. WiFi and wireless networks have specific limitations as well, but will not be discussed here. In most hardwired networks, including BAS, these limitations are approximately 2,000 to 4,000 feet. Another issue is that there may be a limitation to the number of nodes that can be supported by such a network. In addition, it is becoming more common to mix protocols within a given system. This limitation requires that special hardware be added to hardwired networks. There are several types of devices that have been implemented with networks to address these issues including repeaters, bridges, and gateways.

Repeaters are used to amplify and regenerate signals for the network, and this generally enables the signal to travel longer distances on the path. Boosting the signal through the repeaters is also used at times to increase the number of nodes that the network can support. This is important for nearly all networks, but particularly BAS because of the number of controllers that are being added to networks.

Two other devices that have been used extensively in the IT industry are bridges and gateways. The concept of an inter-protocol network is that more than one communication protocol is implemented within the same system. Bridges are used to provide inter-protocol functions when the two networks use the same or similar protocols. This may occur when multiple generations of the same product are installed in the same system. Another example is when the system is expanded beyond its node capacity, and the bridge is used as a way to combine two networks that employ the same protocol, but are operating independently.

✓ Gateways are extremely important to many of the BAS, energy and other related products that will inevitably be used to provide data for energy and analytics. There have been many gateway products introduced to the energy and buildings market, and these devices have not always offered flawless operations, primarily because they typically require specialized knowledge and ongoing maintenance and management for optimum performance.\Gateways are employed when communication is required between protocols that are totally incompatible. Middleware, as will be discussed in a later chapter, provides many of the functions that are discussed here, but on a larger scale and with more capability.

Gateways have been used with many conventional BAS systems from a single manufacturer. This is particularly true given the extensive number of mergers and acquisitions within this industry, and the need for a manufacturer to integrate these desperate products under the same corporate banner. Equally common is the use of gateway products to extract data from meters, computerized maintenance management systems (CMMS) and a host of other technologies. Another common application for gateways has been to integrate BAS and various distributed processing systems, which require interaction between distinct functions such as analytics for enterprise energy management (EEM) or fault diagnostics and detection (FD&D), as well as building automation for HVAC, life safety and security systems. Gateways have also been used to implement web-based dashboards and other software products. In essence, the gateway provides a data conversion process in software. It is important to note that a gateway may be an independent piece of hardware and software, but it may also be a software function that is integrated with any piece of BAS or other equipment, such as an enterprise supervisor. This may seem confusing but the optimum implementation for any market participant will vary. Ultimately the function is to handle requests for data from a dashboard, or other interface, and then to translate that request and create a message that the target system understands and transmit it. The data are then received by the gateway in the form of a message and are then converted for the dashboard, etc. and transmitted for display through a user interface. In many cases the gateway is completely transparent to the user, yet this does require that protocol be acquired by the software developer.

This discussion of the buzzwords associated with data communication, the OSI model, network hardware and software is admittedly brief. The intent is to provide a basis to better understand the role of these topics for energy and analytics. To expand on these concepts and provide the reader with a more focused view of the issues at hand, this chapter will now briefly discuss BAS industry networks and communication. BAS technology is not the only source of data for analytics technology, but it can provide significant amounts of data, and BAS networking and data communication are also representative of the approaches that are used for many metering, CMMS and other systems that are also used as sources for data. Therefore it is of interest for the reader to consider this content as a good illustration of the approach taken to access data without having to install vast networks of new sensors. Leveraging data communications and tools like gateways enables technologists to overcome the high cost of sensor networks during the implementation of technology for energy and analytics.


BAS INDUSTRY NETWORKING

This section is key to the discussion of BAS networking. Through the American Society of Heating Refrigeration and Air Conditioning Engineers (ASHRAE), and independent efforts of numerous individuals and companies within the BAS industry, there has been a tremendous amount of work done in the area of networking. The purpose of this section is to identify and define some key terms that apply to BAS network protocols. Of course many of these terms are defined the same as with IT networks. The OSI model discussion should prove to be good background, particularly given its broad acceptance within the IT and BAS industry. This section will review a BAS network example with the intent of identifying key terminology for both this specific topic and in a general context as related to IT networks.

**BAS Application Example**
There is a great deal of similarity between network issues for the various industries. BAS however does introduce some new concerns due to the requirement for automation and control. The application

used as an example to identify these concerns will be a medium-sized office building with packaged variable air volume (VAV) equipment. The intent here is not to fully describe the application, rather to identify key aspects of it which are network oriented.

The example application is a 50,000-square-foot office building. Primary use of this building is by commercial tenants who each occupy spaces on one of four floors. HVAC requirements for the building are met with distribution from one packaged VAV air handler. This unit is actually a self-contained VAV unit with 120 tons of cooling capacity. The air handler provides constant temperature air to pressure-independent, fan powered VAV terminal units or boxes. The VAV boxes are fitted with integral electric reheat to meet building heating needs. There are approximately 100 VAV boxes in the building with roughly 25 per floor.

**Control Equipment**

Control for this building is provided through a BAS with intelligent microprocessor based direct digital controllers. The BAS devices are integrated with the air handler, each VAV box, etc. Each piece of equipment is fitted with a self-contained device and includes network hardware and software components. BAS control devices of this type are capable of maintaining a control strategy, or algorithm, for any given piece of equipment. Control may include temperature, pressure, humidity and other strategies along with equipment safety monitoring for alarm or failure conditions.

The air handler (VAV-AH) is fitted with a BAS equipment level control to provide discharge air temperature and reset, along with duct static pressure, control. This BAS device may also monitor compressors for safety conditions, such as high or low pressure.

The VAV boxes are fitted with zone level BAS devices to control fans (if applicable), space temperature and pressure control in some applications. Pressure independent VAV boxes are able to integrate zone damper control with an input that monitors main supply pressure. In this way it is possible to maintain a more consistent CFM supply to the space for comfort and also to reduce noise in the building.

Many VAV-AH equipment level controllers are also capable of integrating zone data for control. The integration may be done through a higher level control or via VAV-AH communication capability on the network. An example of this would be integrating zone data with VAV-AH control functions for reset of discharge air temperature and pressure data for building pressure control. Building pressure allows a manager to operate all or part of a building at positive or negative pressure. For example, a lab could be controlled to negative pressure to avoid introducing undesirable contaminants into the rest of the building.

The final area of control is lighting for common areas which is handled by switching electrical panel circuits. Lighting may be controlled on a time-of-day basis and may be overridden along with the HVAC for after-hours tenant use. Local sensors are often integrated for occupancy or daylight harvesting, and it is possible to integrate building-wide algorithms for demand response and more extensive scheduling and sustainability routines.

## Control Network Overview

Network communication, almost exclusively utilizing the BACnet™ standard for BAS, enables data communication between all devices. This would include, for example, sharing data between the equipment level device on the VAV-AH and the zone level device on VAV boxes. This network also serves to provide VAV-AH data to higher level controllers and allow commissioning as well as interrogation of controllers on the network. Of course this network may be the source of extensive amounts of data for FD&D, EEM and other analytics functionality. The most common physical layer media for such a BAS architecture was traditionally twisted pair wiring, but modern systems utilize wireless, fiber optics and any number of different media.

## General Control Network Communication

It is important to discuss network communication capabilities in the context of a BAS architecture. As noted in Chapter 10, architecture refers to the rules that define the functions and capabilities of various BA$^{DDC}$ devices, and these rules are established by the protocol in use. The goal here is to help the reader gain an understanding of the varying levels of complexity found in BAS networks. BAS network architectures allow three types of network communication: network interface, distributed network data, and integrated control. Communication types are associated with the level of controller sophistication. Network interface is a minimum requirement for any distributed network, while the other functions tend to be more powerful features. This is also the most important requirement to provide data access for energy and analytics tools. Nearly all BAS will make provisions for all three communication types, however lower end systems may be very limited in their implementation.

Network interface is critical for access to all controllers on the network. Access may be through a local terminal or web-based dashboard. In addition to access to BAS data, this interface also provides a means for commissioning a site, and all the associated controllers.

There are two types of distributed network data (DND), global and shared. These data types were discussed in Chapter 10 under the BAS. Integrated control occurs when network control commands consist of control decisions made in one BAS device and carried out by one or more other BAS devices. Integrated control is important for system-wide functions like critical alarm shutdowns, demand response (DR) or demand limit control (DLC). To implement DR, a network device must be identified to monitor the internet for a DR event notification. For DLC, one of the BAS devices will monitor the utility demand meter and determine when a pre-determined peak is imminent. When the DR event or DLC peak occurs, the BAS will execute a sequence. The intent is the same in both cases, to take action that reduces load. In the case of DR, the actions may be different and will have been agreed to between the customer and the utility. With DLC the actions may be the same or different, but rather than taking action on the event, the goal is to issue commands rapidly to the BAS devices and ensure that the demand setpoint is not exceeded. DLC commands along with other integrated control network functions must also be transmitted quickly, usually within a frequency of seconds.

This brief overview of BAS networking technology is provided to assist the reader with an understanding of one of the more common systems that provides data to analytics. Of interest is that the same approaches would be found in metering and other systems. Ultimately however, the point is that energy and analytics can be implemented in the most cost effective manner by leveraging data points from BAS and other systems, therefore it is in the reader's best interest to have a working knowledge of how such systems operate.

## INTERNET AND WEB SERVICES

The previous discussion focused on BAS networks and was intended to provide a general understanding of that technology. Those systems might be referred to as "intranets," however during that discussion there were numerous references to web-enabled dashboards and the internet as a vehicle for transporting BAS data to analytics suites. It is therefore appropriate to provide a brief introduction to the internet and web services here as well. The next chapter will also provide more content on this topic. This introduction is intended to expose the reader to the next layer of architecture—worldwide architecture via the internet.

It might be helpful to begin with a discussion of web services, because these services are being used to enable data access. It may also be helpful to consider definitions for web services from IT industry powerhouses. According to **IBM Corporation** "Web services are self-contained, modular applications that can be described, published, located, and invoked over a network, generally, the World Wide Web." **Microsoft Corporation's** description is more succinct, "a web service is programmable application logic, accessible using standard internet protocols."

Others define web services as a business logic or as information made available using the XML (extensible markup language)-based SOAP (simple object access protocol). In the analytics space, however there has been a migration to other technologies such as representational state transfer (REST) and other models, as many believe these are easier to implement. REST will be discussed in Chapter 12 on middleware. In any case, web services are invoked to access data from a variety of intranet and internet data sources. To accomplish such tasks, technologies and protocols have been developed that can integrate existing business processes and resources, as well as systems data from BAS, etc., and make them available over the web. Energy and analytics tool developers looking for approaches to develop web services, have new integrated environments to choose from that offer everything from web servers to application development tools.

The BAS and analytics industries have become very aware of these trends. The site www.automatedbuildings.com has provided an incredible source of data for the BAS industry, but equally important has also become a clearinghouse for data on a wide range of emerging markets including energy and analytics. The following article is excerpted from a January, 2013 issue of AutomatedBuildings.com and is called **Information Model: The Key to Integration**, by Eric Craton, Product Development and Dave Robin, Software Development. The authors state that web services are a new breed of web application. web services are self-contained, modular applications that can be run over the internet and can be integrated into other applications. Web services perform functions that can be anything from simple requests to complicated business processes. For example, a weather bureau could offer a web service that allows a BAS to automatically retrieve temperature forecast data for use by various control algorithms. Similarly, an analytics suite can use web services to extract data from BAS and other systems to be used for analysis of critical energy and building functions. Equally important, the best analytics tools analyze that data in the context of business process, because interpretation of the data and analysis results requires a detailed understanding of the process being executed in the building. For example an analytics tool could utilize web services to extract data from a tenant's accounting system to obtain up-to-the-minute figures on energy consumption, and conduct analysis that evaluates the business process against energy consumption data to determine if changes can be made in operations. In the past, this type of data exchange would require a custom, "hard coded" data request to retrieve information that already existed in the host computer. A web service, on the other hand, is a way to allow any authorized client to actually run an application on a server in the cloud to access data and conduct analysis that previously was not possible. This provides a mechanism for building owners to not only conduct the analysis but be more timely to acting on recommendations to improve the operations. Since BACnet objects are information models, and XML is a modeling language, we could express these high level information models in XML and in so doing make them compatible with the emerging web services architecture.

## Architecture

This chapter has covered data communications, networking and protocols. It also provided some background on BAS networking, with BAS being representative of many systems that are to be found in operations with energy and buildings. The term SCADA, system control and data acquisition has not be used, but SCADA is also common in this arena and is similar to BAS. This last section in the chapter will revisit the topic of architecture. As discussed there are a large number of rules or standards, which are collectively called data communications protocol. Protocol is an essential element in defining the capabilities and characteristics of a network and in defining the network's architecture. The concept of architecture is defined by both the protocol and network software, which was discussed earlier in this chapter. Architecture encompasses a broader scope because it involves both the physical and network implementation of the protocol. This section covers the physical implementation of architecture through a discussion of topology. Topology defines the structures that are possible and also the physical wire or media that may be used in connecting the network members. Topology is defined by standards that are established in layers one (physical) and two (data link) of the OSI model. Network implementation is achieved through the interaction of the protocol. The architecture allows communication between the various network members. International network standards have been developed to define architectures. These standards along with the interaction noted above, dictate the physical media and structures as well as the process for passing data between network members or nodes. Networking, the support of communication between distributed intelligent controllers, is essential to establishing a reliable flow of data for analytics. Networking is addressed here as one of two essential elements of an architecture, in terms of technology available and standards which are applicable in the controls industry. The central focus will be on understanding the interrelationships within the architecture and roles of both critical elements, protocol and network software. This will be useful because a solution to the protocol issue requires combined implementation of a standard protocol within a standard architecture. The term architecture will be used to describe the complete solution which includes the elements of protocol and topology. Network software is also of interest under the topic of architecture but this topic was covered previously with the discussion of IEEE standards for media access control.

With that introduction it becomes obvious that the concept of protocol should be transparent to the user, though all of the aspects discussed under architecture require implementation of a common protocol. So the focus here is on implementation of the complete architecture, including protocol, but discussions will be limited to topology as implemented by the protocol and network software as it interacts with the protocol.

To aid in understanding architecture, it is helpful to address the specific physical connection of network members, and the implementation of the protocol messaging through network software. This should answer these questions: What does the network look like, and how is it physically implemented? Then the discussion will view the interaction between members of the network. To clarify the residence of these components however, the reader should understand that all of the questions posed above are answered in one of the lower protocol layers. The more complex user-oriented interactions occur between the upper layers of the protocol and the network application software.

Rather than covering these topics in relationship to the specific layers in which they reside, they will be covered in conceptual terms. The intent is to cover the concepts in a way that is consistent with how system professionals view a system. The general architecture will be discussed first, followed by an overview of the physical layout or topology, and finally a review of the software applied within the architecture. The umbrella term for all of these discussions is "architecture," and it will be used in conjunction with "topology" and "network" to describe the characteristics of implementation. Again the content here is divided into two categories, architecture and topology.

It is not uncommon for these terms to be used interchangeably, so it is important here to define them very clearly. Architecture, as noted, is the overall definition of a network including protocol, topology or physical wiring connections, network implementation or rules of data link control network application software. Key aspects and capabilities of communicating controllers will be discussed within each category.

ARCHITECTURE

This section will focus on the two key elements of architecture; topology and networking software. Architecture is a much used term, and interpretations change radically based upon the context, so a definition is useful. An architecture is defined as the protocol (including electrical signaling via topology and if necessary media access standards) and networking software that allow interconnection and communication with building control system(s) and their associated components. Physical and networking standards that combine to define the architecture, as outlined by Uyless Black*,

---

*Uyless D. Black, *Data Communications and Distributed Networks*, Second Edition, Prentice Hall, Inc., Englewood Cliffs, NJ, 1987

include: hardware, software, data link controls (DLCs), standards, topologies and protocols. Communicating for IT, BAS or any other type of system began when more than one panel with intelligence was applied in an architecture. For example, the most common BAS industry architecture has been a master slave implementation in which one device is responsible for managing communication. This device more than likely contained all of the control algorithms, as well as communication responsibility. The most common approach was to implement slave panels to provide inputs and outputs, thus expanding the capability of the system.

Architecture describes the overall communication systems for any network including BAS. Within any architecture there may be a variety of devices, and among those may be specific devices that have certain capabilities for communication and interaction with other network members. A term the reader may have heard is "level of architecture." A level of architecture implies a system, or portion of a system, in which all devices communicate on a common basis. It is also possible for an architecture in this context to encompass multiple protocols. As noted above and elsewhere in this text, various styles of interconnection are possible that allow complex protocol interaction. For this chapter however, the assumption will be that the architecture shares a common protocol. This will allow the author to focus on the most important concepts of architecture to our target discussion of topology and networking software.

TOPOLOGY

Topology is defined by the 1st and 2nd OSI model layers and establishes the structures that are possible for interconnecting nodes. Topology also establishes the physical wire or media that may be used in connecting the network members. In addition to defining the rules for network interconnection, Topology also covers the associated characteristics of that connection. For that connection to occur there must be rules for attachment of each node to the local area network (LAN), and for the overall structures that are acceptable for multiple nodes. Per Uyless Black, there are four major components that define a local area network.

1.    Communication channel

2.    Interface

3.    Protocol control

4.    Intelligent device (i.e. enterprise supervisor or dashboard)

The communication channel or path is critical to a network. The path is defined by a requirement for compatibility with both the protocol and the network software, and varies significantly by the implementation. Path media ranges from the simplest twisted pairs of wire, called "twisted pair," which is low in cost and easy to install, to fiber optics with higher data transmission capacity and speed capabilities. Again the path in most cases will be defined by the lower layers of the protocol, and is the essential component of the topology. It is also relevant here to include a comparison between topology and local area network (LAN) type. LAN type is a noteworthy because it has a dramatic impact on overall network performance. Two LAN types are to be discussed: broadband and baseband. This refers, in simple terms, to the capacity of the physical media to carry data, and may also determine the physical media that is compatible. The only point here is that the amount of data that are being transmitted from any system to an analytics suite may require consideration of a LAN type that can support the desired bandwidth. The interface between the path and the protocol logic is the second key component of this discussion. Again this interface is defined by the protocol and network software or operating system. The interface is typically a software component. The orientation varies, yet in essence the interface must allow for an electrical signal from the communication path to be converted to the upper layers of the protocol. This is a good example of a data link layer function. This interface is software that interacts directly with the physical layer and its associated hardware.

Regarding topology, there are a number of options for organizing control networks including those listed below. This level of detail may be more than is necessary for readers to understand as they develop strategies to access data for analytics, so the content will be covered at a fairly general level. It may also be helpful to differentiate simple concepts from more complex topology approaches that are applied with network technology. These topics will be discussed in two categories: simple networks and network topologies.

Simple Networks
- Point-to-point
- Multipoint

Network Topologies
- Star
- Ring
- Bus
- Hierarchical

Ancillary LAN devices that directly impact the network include: repeater, gateway and bridge which were previously discussed. LANs are covered because they often define the media that is used to connect the structures noted above, and include broadband and baseband.

*Simple Networks*

Simple networks are oriented to IT or computer systems but provide good background for the discussion of network topologies. In this case the IT network may be enabling communication between servers or devices equivalent to an enterprise supervisor, but for simplicity this section will refer to the device as a computer. Local area computer networks are oriented to limited geographic areas and with simple networks the assumption is that only one computer is present. That computer typically supports communication to allow interface from several terminal or workstations.

The point-to-point network style is extremely simple and tends to be very much oriented to computer networks. The structure includes one computer, a communication path and a dashboard or interface of some type. This point-to-point approach barely passes for networking, yet it is considered simple networking because the terminal uses a protocol for either direct connect interaction or communication via an IP address to the internet.

The multipoint network architecture is an expansion of the point-to-point approach in which more than one interface, display or dashboard is possible. The multipoint network was developed to allow more than one interface to be connected simultaneously to a single computer or server. This is more typical with computer local area networks (LANs) that require interaction from interface terminals with the server for access to application software and a database.

*Network Topologies*

Network topologies are distinguished from simple networks by their more specific orientation to networks with multiple intelligent devices or servers. The star network is common with centralized computer networks. Typically these applications involve primary computing that is contained in one site, and all of the remote interfaces are connected to that site. Again interfaces in this case are more likely to be computers or dashboards. Two of the most popular local area networks (LANs) in the computer industry, ARCNET developed by Datapoint Corporation and Token-Ring by International Business Machines Inc. (IBM), allow the star network topology. The star topology enables a network which allows for a central hub and device wiring runs that can terminate at the last device. The ring topology requires that all nodes be connected to the physical media in a closed loop. With the ring concept each node is linked to the adjacent nodes on both the left and the right. Again it is common to apply ring networks in the IT industry, and in fact the IBM product noted above is a good example of a ring network. Some benefits of the ring concept are that higher data transmission speeds are possible and data collision schemes can be implemented in a simple fashion. With a ring network each station in the ring receives and retransmits a message, and as a result data collisions and errors can be minimized. There are some detriments to the concept which should be considered. First, rings have less flexibility in terms of installation, and further may add significant cost to a project due to the requirement for completion of a closed wiring circuit. Another consideration may also be the capability required of each controller node for receipt and retransmission of data.

The bus network structure is distinguished by the central "backbone" which is the physical media. The media also allows taps for arms or branches, each typically allowing one controller to be connected. The bus topology is extremely common with BAS networks because it lends itself to single wire runs without the need to loop back, as in the ring network. Generally the physical media in control bus networks is a twisted pair, which may be run throughout a building. The communication signal travels along the bus, and every controller node "listens" for the signal that carries its address. In most cases the controller both receives and transmits via the same twisted pair bus using a special signaling process. Both bus and star systems require that a "biasing resistor" or other component be used to terminate the bus. This termination is what allows the installer to establish an end point for data transmitted on the network. These topology structures have been outlined briefly for the reader's information. An interesting note is that networks are not limited to a particular topology, sometimes these approaches may coexist.

LAN type is the next topic, but it is complex, and need not be understood for analytics, so it will not be covered in detail. LAN types, broadband or baseband, also do not have a major impact on the architecture. This is a physical media issue because data are transmitted over the physical path via an electrical signal. The reader is likely aware that all computer data are, at their most basic level, zeros or ones. These zeros and ones are converted to electrical signals for transmission over the physical media. In simple terms, the greater the span of frequency that a path can support, the more information that may be transmitted at one time. The only important note for the reader regarding energy and analytics is that frequency and bandwidth are directly related. For illustration purposes, consider the analogy of an interstate freeway through a major metropolitan area. The more lanes of traffic that the freeway can accommodate, the greater the flow of automobiles in each direction, and therefore the system offers better "throughput." Throughput in data communication, as the term implies, is the measure of information traveling on the physical media for both transmission and response. In data communications networks, the concept of throughput is also impacted because greater frequency is typically accompanied by techniques for increasing transmission speed.

The concept of a computer LAN is to allow multiple devices, from small tablets and smart phones to personal computers, laptops and servers to share application software and information. A well-defined means must be established for those nodes to gain access to the media, and for recovery in the event of data corruption. This is accomplished by network software that was discussed earlier.

Data transmission media, speed and limited processing requirements at each node of a LAN are also required.

## ENERGY AND ANALYTICS DATA ACCESS

This text is focused on energy and analytics as applied in a wide range of applications from buildings, plants and large-scale campus or city operations. In the end there will be a host of systems and technology already in place that can be rich sources of data for this analysis. The intent of this chapter has been to provide a limited overview of the data communications and networking that is deployed by typical systems of this type, and provide a limited understanding of the techniques that would be required to access the data in those systems. This access to system information will typically involve interface via the internet and a means of initiating data requests via an enterprise supervisor that has access. This process must be transparent to an analytics engine, allowing for a request of status on any individual point or group of points. Equally important, this process must be resilient to ensure that the flow of data is continuous, reliable and accurate.

# Chapter 12

# Middleware—A New Frontier for Building Systems and Analytics

This may be one of the most important chapters in this book for the long-term success of energy and analytics. With the foundation that was provided in the prior two chapters it is possible to peel a layer or two from the onion, and to begin to understand the underlying technology, or glue, that enables building systems to interoperate. Glue is a good word to describe "middleware" because these components actually provide connective tissue between systems, but a better term might be "infrastructure." The infrastructure of energy and analytics is made up of systems with hardware devices, networking and interface software and an infinite number of instrumentation devices (sensors, meters, etc.). All of these system components must be accessed by and communicate with an analytics "engine" or "tool," which conducts all of the analyses. The purpose of this chapter is to provide insight and information to the reader that will be invaluable in understanding how these systems operate, and will be useful in comparing analytics tools and system approaches between various suppliers.

Perhaps one of the most confusing things about this topic is that words like middleware are not familiar to many professionals in the energy and buildings industry. In addition, middleware, as a term, falls short of describing the full functionality of this technology. However, the industry embraces certain terms as identifiers for groups of products and services, and once a term sticks, it becomes useful to identify the technology being discussed. So the overarching term for products that will be discussed is middleware, but this simply provides a starting point, and again the goal is to provide an understanding of the underlying and interconnecting technology that makes it possible to achieve "interoperability." It may be most useful however to provide some history from the energy and buildings industry to help clarify why this functionality is needed.

In the 1980s a movement arose within the building automation systems world that demanded standardization of data communications. The term "open systems" eventually caught on as a way to describe what users wanted. Whether control technologies are referred to as building automations systems (BAS), energy management systems (EMS) or direct digital control systems (DDC), this technology performs the same work and will simply be called BAS here. Chapter 10 referred to BA$^{DDC}$ because the emphasis was on the automation and control functionality itself. Here the reference will be BAS because the focus is on the systems themselves and the technologies that enable data sharing and interoperability between them. So BAS provides a focal point for this discussion. By the mid-1980s a number of market drivers, i.e., energy crises, development of microprocessors and demand for energy management, had resulted in hundreds of millions of dollars in BAS installations. As managers, particularly those with multiple buildings, or those who operated complex and ever changing campuses, became aware, the generation of BAS at that time presented a number of communication issues. These early-generation BAS were called "legacy systems." Recognizing the limitations and challenges presented by legacy systems led building owners and managers to drive the open systems movement for two reasons: 1) the absence of standards required independent computer interfaces for every manufacturer's BAS, and 2) expansion of systems was difficult because BAS from multiple manufacturers could not interoperate, which made owners feel locked in to a particular system. The rest is history, and by 1995 the BACnet™ had been developed, and in another decade products deploying that standard became ubiquitous. Over the period of three or more decades however, managers were still demanding solutions for those hundreds of millions in legacy systems, and that is how "middleware" came to be.

MIDDLEWARE

Middleware, as a term, was borrowed from the information technology (IT) industry, in part because of the types of devices that were used to apply the functionality. According to Wikipedia, "Middleware is computer software that provides services to software applications beyond those available from the operating system. It can be described as 'software glue.' Middleware makes it easier for software developers to enable communication between multiple systems and to access inputs/ outputs (I/O). This allows developers to focus on the specific purpose of the application being developed. Middleware is the software that connects software components or enterprise applications. Middleware is the software layer that lies between the operating system and the applications on each side of a distributed computer network."

It may be clear why the term stuck with BAS because the desire to unify systems interfaces into what we later began to call "dashboards" required software to perform communications and access input/output and other data.

The need for a type of software/hardware, which is referred to here as middleware, became even more pronounced as managers ended up with multiple generations of product from multiple manufacturers. A unifying middleware product that allowed managers to interface with all of those systems through one PC or dashboard was highly desirable. Setting aside the actual interface for a moment, one of the ways that this functionality was accomplished initially was via a gateway. In Chapter 11 gateways were described as pieces of hardware that allow the same network to have more than one protocol. The BAS task entails a bit more complex functionality, but ultimately these "BAS Gateways" were developed as pieces of hardware that could convert a legacy BAS protocol for communication over a larger network to a higher level computer, and ultimately a dashboard. Variations on this same approach are still being used today to provide interoperability that enables data to be accessed from various systems and made available to energy and analytics tools.

In 2008 Cisco Systems teamed up with Frost & Sullivan to publish a white paper called "Intelligent Middleware." Cisco had acquired Richards-Zeta, a progressive building automation company, and their platform was called the Mediator. A major focus of that paper was migration of building automation systems toward internet protocol (IP) integration. Figure 12-1 comes from that white paper, and provides a visual to emphasize Cicso's position that middleware technology was critical to allow interchange between BASs-using proprietary protocols and IT centric systems using IT protocols. The notion was that a "capability gap" existed between BAS, which were not capable of functionality beyond their intended purpose, and the internet with more sophisticated functions for a larger enterprise. Today, energy and analytics would be one of those more sophisticated, or larger enterprise, functions, but Figure 12-1 predates the emergence of analytics as it is now known. This

# The Mediator – Enabling Convergence from the South to the North

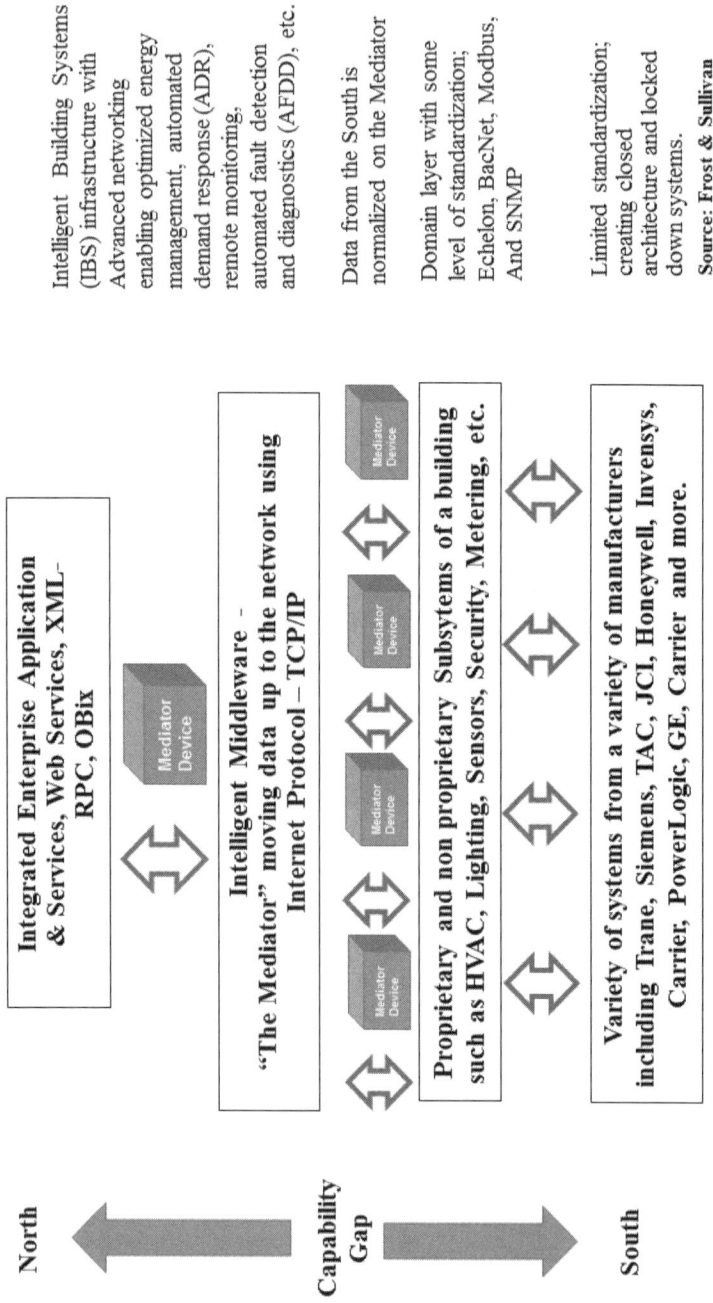

**North**

**Integrated Enterprise Application & Services, Web Services, XML-RPC, OBix**

Mediator Device

**Intelligent Middleware –
"The Mediator" moving data up to the network using Internet Protocol – TCP/IP**

Mediator Device   Mediator Device   Mediator Device

**Proprietary and non proprietary Subsytems of a building such as HVAC, Lighting, Sensors, Security, Metering, etc.**

**Variety of systems from a variety of manufacturers including Trane, Siemens, TAC, JCI, Honeywell, Invensys, Carrier, PowerLogic, GE, Carrier and more.**

**Capability Gap**

**South**

Intelligent Building Systems (IBS) infrastructure with Advanced networking enabling optimized energy management, automated demand response (ADR), remote monitoring, automated fault detection and diagnostics (AFDD), etc.

Data from the South is normalized on the Mediator

Domain layer with some level of standardization; Echelon, BacNet, Modbus, And SNMP

Limited standardization; creating closed architecture and locked down systems.

**Source: Frost & Sullivan**

Figure 12-1

figure assumes that the mediator provides interface to older-generation BAS-using proprietary protocols only, which have been called legacy systems in this book. However, today the concept for middleware is more expansive. In addition to legacy systems, it is not uncommon for middleware to provide interface between many systems using proprietary protocols as well as BACnet and other standards. Of equal interest, these middleware devices may also integrate an instantiation of energy and analytics functionality as well. The analysis is still occurring in the cloud, but more companies are developing their own energy and analytics tools and using middleware as a focal point for the solution. Therefore with analytics, the line sometimes blurs between *traditional middleware functionality* (providing "drivers" to execute data conversion between proprietary standards, open standards and IP), and *middleware for buildings* (providing traditional functions plus added functionality that is central to a product offering, such as enterprise energy management or analytics).

Also around 2008, Ken Sinclair, publisher of Automatedbuildings. com, authored an article on middleware. Ken began with common definitions for middleware, that it is "the 'glue' between software components, or between software and the network, or it is the slash in client/ server. Middleware is used to connect applications to other applications. (See Figure 12-2.)

Middleware opens opportunities to use multiple user interfaces, such as mobile dashboards running on tablets, phones and other web-enabled mobile devices, to monitor and interact with different systems. This (middleware) brings together data from multiple systems, in multiple buildings, to give facility managers unprecedented control over their energy use, help them spot trouble with equipment early, and enable them to make adjustments in real time. In the past, it was common for this middleware to be packaged with the automation system, security and life safety systems from one vendor. While this was convenient for the purchaser, it left no room for future enhancements from other vendors—they were stuck with the vendor that installed the system. This has changed in recent years, with many vendors using open protocols, or standards, which allow customers choice. The bigger challenge today is determining what information and systems should be integrated at the building level. You can't manage what you don't measure, but careful choices have to be made regarding what sort of level you both need and want."

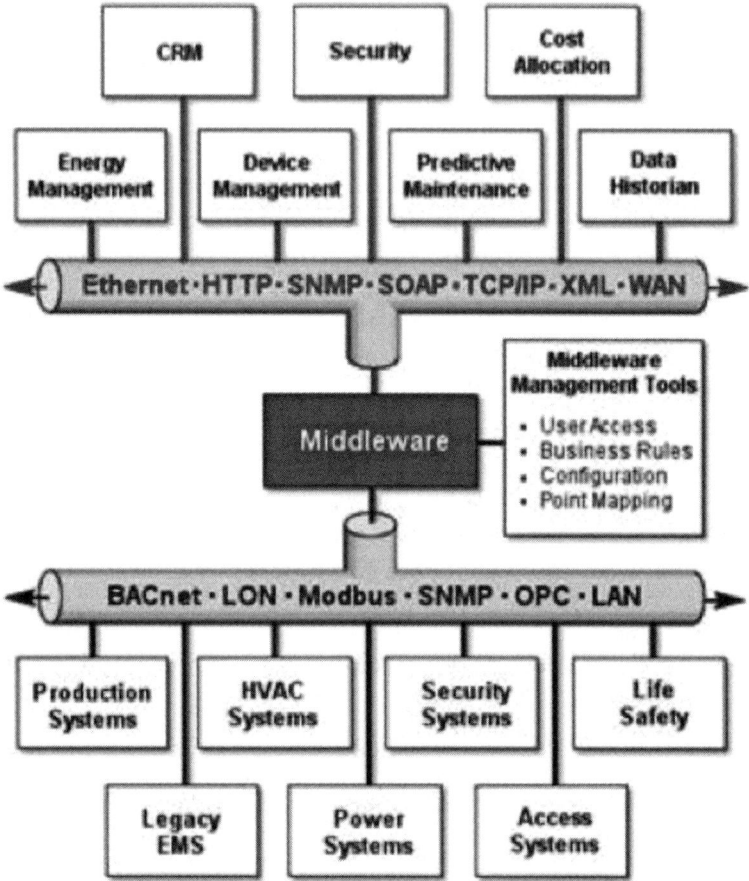

**Figure 12-2**

Another take on defining middleware came from Jim Sinopoli and Neil Gifford of Smart Buildings. In a 2008 article (www.automatedbuildings.com) they said that the implementation of a middleware solution would normalize and standardize the data of building automation systems. In addition, these systems can extract and digest the data and control the systems, such as set device values or set points. The word "middleware" may sound like just software, but their implementation requires software and hardware. The hardware will involve subsystem controllers likely needed for the HVAC systems and some IT hardware, such as gateways for interface to servers and dashboards. Figure 12-2 came from this article, and it identified four of the companies who dominated the industry at that time. Interestingly, all of the building-specific players mentioned were later acquired: GridLogix, acquired by Johnson Controls Inc.; Richards-Zeta acquired by Cisco; and Tridium acquired by Honeywell. Tridium is a very important participant in this industry, and will be discussed later in this chapter when more detail is provided on middleware technology solutions. Lynx Spring is a Tridium developer and BAS manufacturer and leader in this space.

Standards are a key topic that has been discussed throughout the last several chapters on data communications and protocol. Standards are key to this discussion because the more standards that are introduced to the energy and analytics industries, the easier it becomes to access data in a reliable and repeatable fashion. This has been called data resilience. Some of these standards have been discussed at length, such as BACnet™, but others are also important like MODbus™, which is used by many metering systems. Another type of standard is oBIX (Open Building Information eXchange), which is "an industry-wide initiative to define XML- and web services-based mechanisms for building control systems." oBIX is an OASIS standard. OASIS is the Organization for the Advancement of Structured Information Standards, a non-profit consortium that drives the development, convergence and adoption of open standards for the global information society. oBIX is important because its intent is to instrument control systems for the enterprise. "The purpose of oBIX is to enable the mechanical and electrical control systems in buildings to communicate with enterprise applications, and to provide a platform for developing new classes of applications that integrate control systems with other enterprise functions." The point with this standard was to overlay an IT solution that can access data in a standardized format regardless of the machine communications protocols. oBIX was ideally targeted to address the data access needs for energy and analytics, however a common theme with technology is that change is constant. Later in this chapter there will also be discussion of web services technologies that are enabling analytics developers to simplify and fast track solutions. Besides these efforts and oBIX, there are other groups working to unify data communication standards and to approach data access in different ways as well. Among these efforts are the work of the Smart Grip Interoperability Panel (SGIP) www.sgip.org, which is a spinoff of the U.S. Department of Energy Smart Grid work. This is a great example of how BAS technology has become part of much larger standards efforts and is benefiting from the work of a much larger community. This author is chairman emeritus of the DOE Gridwise Architecture Council, and represented the energy management and buildings domain with that group, but there were wide ranging other domains represented as well: industrial automation, electricity infrastructure, residential systems, IT, networking and more. An important milestone during this author's term as chairman was passage of the Energy Independence and Security Act by the United States Congress. That law directed the National Institute of Standards and Technology (NIST), which is part of the U.S. Department of Commerce) to lead a national effort to identify technology standards that could enable a Smart Grid. Interestingly, Steve Bushby was one of the managers at NIST who oversaw this work. Mr. Bushby brought strong buildings knowledge as well, because he was a founding member, and one of the leaders, of the BACnet™ standards committee with the American Society of Heating Refrigeration and Air conditioning Engineers (ASHRAE). The SGIP is actively engaged in a wide range of efforts to identify technologies and standards, which are likely to prove useful in data access for energy and analytics. Another effort that will be discussed at greater length in a future chapter is Project Haystack, an analytics industry 501.C3 organization engaged in an open source initiative to streamline working with data from the Internet of Things. Haystack is standardizing semantic data models with the goal of making it easier to unlock value from the vast quantity of data being generated by the smart devices that permeate homes, buildings, factories, and cities. Applications include automation, control, energy, HVAC, lighting, and other environmental systems.

Of particular importance to the topic of middleware is data communications and networking. The networks however are not just Ethernet or other IT networks; there are a host of network technologies in play. Therefore, the next topic is to define a number of these networks that are being managed and utilized via this middleware to make data available to analytics.

NETWORKS FOR ANALYTICS

The topic of networks was discussed at length in Chapter 11. This section may repeat some content, but the goal is to build on this topic. Again a local area network (LAN) is a system that enables data communication between multiple microprocessor-based computer devices, which must share information to perform work. Protocols and networking software were discussed as the means for enabling communications and data access between different computer devices. The network provides the means to establish communication between multiple microprocessor based devices (servers, personal computers, BAS controllers, etc.), and ensuring that guidelines are established for the exchange of messages containing information or data. The means for sharing that information is a network, and the rules which define the makeup of the data are the basis for a protocol.

With the advent of microprocessor-based systems for building applications—from BAS for HVAC, security, etc. to meter data management and enterprise energy management—data communication and networking became an integral part of the buildings world. With the advent of cloud-based systems including those providing energy and analytics services, this fabric of communication and networking has become even more complex. As discussed in Chapter 11, the local area network (LAN) is a system of computer devices that is interconnected, in a limited area, such as a building, using network media, or physical cabling. LANs for many different types of building technologies including the BAS, have been standardized by the BACnet™ standard. As noted above, there are also a number of other standards, such as Modbus for meters, that are being integrated in a meaningful way through the efforts of the American Society of Heating Refrigeration and Air Conditioning Engineers and the Smart Grid Interoperability Panel (SGIP) to name a few. That integration, particularly with legacy systems that are already installed, is typically enabled by middleware products. This is especially true for energy and analytics, because there is an expansive landscape of systems and technology that may be identified as sources to access data or as enablers to transmit data. This is why the topic of middleware is so critical to energy and analysis, and why the term "glue" seems so appropriate. Middleware technology is used to connect systems that use disparate protocols, and convert the data for presentation to energy and analytics technology. Middleware technology is also used to connect systems that utilize different network technology, and enable its presentation to energy and analytics technology. This section has focused on coverage of LANs, but will now provide some insight into the broader range of technologies that must be interoperable with for energy and analytics including:

Wide area network (WAN),

Virtual private network (VPN),

Wireless networking technologies,

Machine to machine (M2M) technologies, and

Cloud computing.

Wide area networks are applied on a very limited scale for BAS, energy and analytics, yet the concept bears noting because it provides useful context for discussion of networking via the internet, particularly virtual private networks (VPNs). A wide area network (WAN) is a network that covers a larger geographic area, rather than a building, campus, etc., (i.e., any telecommunications network that links across metropolitan, regional, national or international boundaries) using leased telecommunication lines. Business and government entities utilize WANs to relay data among employees, clients, buyers, and suppliers from various geographical locations. The key distinction between WANs and the discussions that will unfold with virtual private networks, etc. is that these networks typically utilize telecommunications systems, unlike VPNs which use the internet. Features of the internet like "always on" and the simplicity of transmitting large volumes of data using a wide range of data formats, as well as a more advantageous cost, have made internet-based solutions, such as a virtual private network, the preferred approach for analytics systems today.

A virtual private network (VPN) extends a private network across a public network, such as the internet. VPNs enable a computer, or a dashboard, to send and receive data across shared or public networks as if they were directly connected to the private network, while benefiting from the functionality, security and management policies of the private network. Quite simply for an energy and analytics application, the VPN provides all the benefits of performing these functions within a building or campus, but it also enables scale. VPNs are created by establishing virtual point-to-point connections through the use of dedicated connections, virtual tunneling protocols, or traffic encryptions. Again this is essential for multi-location building and campus owners, as well as those that have management staff in many locations. The VPN provides a vehicle for transporting the data to all of those locations, where of course the applications can be housed locally on servers. This is particularly important because energy and analytics functionality relies equally as much on human engineering knowledge and skill, as it does on technology. In many cases, the VPN is the communication enabler for the energy and analytics and other applications that are housed in the cloud, but more on that in a moment. The benefit to this is that access to data and analytics applications, from any geographic location, makes it possible to leverage human knowledge to develop a better understanding of analytics results.

The LAN, WAN and VPN as discussed thus far are complete and functional networks encompassing all of the capabilities described under the OSI model discussion in Chapter 11. An important distinction for analytics is that these LANs may incorporate multiple protocols, requiring gateways, etc., which may also requires deployment of a bridge to unify multiple communication media. Middleware is the essential technology to unify all of these differences and create seamless access to data for energy and analytics tools. In the case of VPNs, the media of interest may be a hardwired Ethernet LAN through an IP address, local area wireless technology or "WiFi" allowing an electronic device to connect to the internet. Given the challenges and first cost detriments to installing wired networks for metering and instrumentation devices, as well as the complexity of interconnecting networks where distance or construction materials such as concrete present challenges, wireless has become very popular for interoperability between these networks. Of course, one form of wireless is WiFi, which can then be used for internet access, and has become ubiquitous, enabling individual smart devices (PCs, tablets, smart phones, etc.) to access the internet.

Independent of its use with VPN technology to provide WiFi access to the internet for energy and analytics, wireless technology is also used for local area networking. This focus on wireless LANs, sometimes called WLAN, pertains to a network that has all the same characteristics as any wired LAN, except there is no physical wired connection between sender and receiver. Instead the network is connected by radio waves to maintain communications. There are many wireless networking standards that have gotten significant acceptance in the buildings space, including ZigBee™ and EnOcean™. The key point is that this is simply another form of LAN being deployed. The next major topic area here will be about building communication systems that are made up of multiple networks. Wireless networks are a good example of a "sub-LAN" or secondary network infrastructure that may be used to access data for analytics.

Machine to machine (M2M) is a term that has received more attention recently and has great implications for energy and analytics. M2M refers to multichannel networks made up a large numbers of devices, many of which are small in terms of both processing power and input/output capacity. In the context of analytics, M2M is very important because these devices can provide a significant amount of data from very granular sources. Some examples are meters, motion sensors and photovoltaic panels mounted on telephone poles. M2M is an independent topic, but in many respects it was the precursor to the Internet of Things (IoT), a topic discussed later in the book. Regarding IoT, a November 2014 report from the International Data Center estimates that there will be 30 billion autonomous things (computers, tablets, smart phones, BAS, etc.) connect to the internet by 2020. Those "things" can provide a rich and exciting source of data for analytics.

Cloud computing is the last topic that will be discussed under this heading of networking. This is also a great segue to expand beyond the infrastructure that is used to move data around, to the applications or engines of analysis that are being deployed under the heading of energy and analytics. Cloud computing is computing in which large groups of remote servers, often physically located in data centers or server farms, are networked to allow centralized data storage and online access to computer services or resources. Cloud computing is the term used to describe the use of interconnected business applications over the internet. As depicted in Figure 12-3, the applications are interconnected via web services, and the end user accesses the required service using a web browser or dashboard. Therefore, the application and the infrastructure do not reside on end users' premises. The end user accesses the application on demand, and can concentrate on using the application for its purpose, without investing capital expenditures, thereby avoiding the overhead of installation, networking and maintenance.

Cloud computing is the term used to describe the use of interconnected business applications over the internet. Nirosha Munasinghe, MBusIT BSc BE, a product development manager with Open General in Australia, published a great article on this topic called "The Future of Cloud Connectivity for BAS" (www.automatedbuilidings.com) in June 2010. An excerpt of that article defining cloud computing is provided here. Mr. Munasinghe provided Figure 12-3 to illustrate that the applications are interconnected via web services and the end user accesses the required service using a web browser.

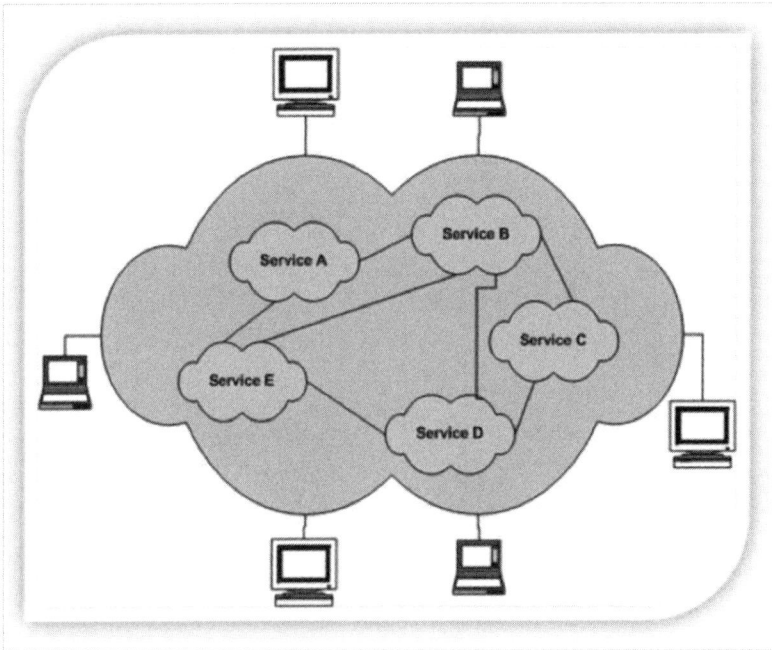

**Figure 3: Set of interconnected business services/applications via web services, communicating with each other is available as a web service or via web browser over the internet for the end user.**

Mr. Munasinghe goes on to explain that each service or cloud consists of detailed architecture and technology invisible to the end user. It consists of the *application/service, platform, infrastructure and storage*, shown in Figure 12-4.

Finally, Mr. Munasinghe briefly describes the functions of each element depicted in Figure 12-4.

- Application/web service: Delivers the software/application as a service (SaaS) to end user or other clouds. For example the end user in BAS can access the user interface for control networks, and it can have services for other parties to obtain data from BAS.

- Infrastructure: Computer infrastructure usually in platform virtualization environment as a service.

- Platform: Allow developers/users to extend or write new applications using the platform as a service (PaaS)

- Storage: Consists of the database to read and write data.

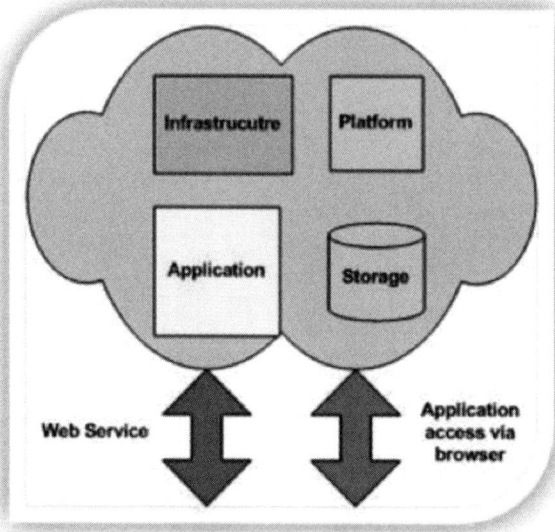

**Figure 12-4: The detail technology behind the cloud**

The early modeling of the cloud computing concept has various deployment methods in *public, private and hybrid* clouds. A public cloud service is available over the internet via direct application or web service to the end user. A private cloud is restricted to private networks that have been referred to in this book as VPNs, for data security and reliability. A hybrid cloud consists of multiple private and/or public clouds to provide enterprise solution. Which deployment model will the building automation market fit into? The vision of smart grid where the BAS communicates with other industries and the need to have a secure control network clearly indicate that BAS market will adopt a hybrid model.

Thus far, middleware has been described as a technology used to form a "glue" that merges systems composed of more than one protocol, as well as more than one network technology. The balance of the chapter will focus on providing some examples of how all of this communication and application technology comes together to be deployed as energy and analytics solutions. In that context it should be helpful to provide a limited outline of technology platforms that are used to deliver this technology. The chapter will close with a brief discussion of web services and related technologies that make up energy and analytics solutions, along with two important topics for this discipline, and building technologies as a whole—network resilience and cyber security.

## ENERGY AND ANALYTICS NETWORK SYSTEMS

Professionals in the energy and buildings space have become very familiar with the idea of complex networks and topologies within building automation systems (BAS). Initial BAS technology was developed with an approach that might be called a "unified network system by design" (UNSD). To truly establish a basis for the complexity of this discussion, it is only fair to say than the notion of a UNSD is relied upon for many more systems than the BAS. It is relied upon for the entire universe of technologies, from meters to CMMS to security for access control, etc., fire systems and many more. There are some companies within the analytics space that have tried to provide comprehensive solutions to incorporate many of these systems, but the business model is extremely difficult. In the simplest terms, the reason for this has to do with the value proposition. With the exception of customers that have highly complex operational environments, such as universities, hospitals and manufacturing plants, no customer has seen value in comprehensive services. The value proposition that has been envisioned by some providers must support the cost of deploying a comprehensive system, spanning numerous independent specialty technologies, and implementing a complete new network of instrumentation along with energy and analytics software. However, there are a handful of companies which have created a compelling value proposition, using middleware solutions to create an *ad hoc* network of existing technology that may be used to provide access to comprehensive datasets from multiple existing systems, and make them available to the analytics engine or tool. The fact that this can be achieved is why the topic of middleware is so important to energy and analytics. Yet such solutions require energy and analytics providers to have sophisticated skillsets, including deep technical knowledge of networking and data communications as well as analytics, to ensure the long-term viability of the implementation.

Two terms from the previous paragraph require a bit more discussion before moving on to discuss a systems example: UNSD and *ad hoc* network. The concept of a unified network system by design (UNSD) is straightforward and should not require much discussion. It is obvious that in the case of UNSDs a company develops a complete system incorporating hardware and software. The UNSD will include an operating system and software to carry out specific functions, and will also include the data communications, networking and web services necessary to provide interface to a fully functioning system. Again this is obvious, and not unlike services that might be expected from any computer-based system manufacturer, but it is important to emphasize that with an *ad hoc* system there is no "company" with overall responsibility to make sure the system operates. More importantly, there is no company to ensure that the system continues to operate under changing conditions as well. Changing conditions, in this case, could be anything from internal network management and IP addressing schemes for any LAN to software and operating systems compatibilities, or could simply refer to general maintenance of the data communication aspects related to access the information. This is not to say that a solution provider would not attempt to provide this overarching service, but the point is that the provider does not have control over the ever-changing conditions, nor over how their technology might operate after those changes occur.

The topic of middleware really focuses on *ad hoc* networks, and this again is because it allows systems to be deployed with lower upfront cost by avoiding installation of extensive sensor and instrumentation networks. Middleware achieves this by opening up access to existing sensor and instrumentation networks through communications technology, but this requires *ad hoc* networks. So what exactly is an *ad hoc* network? *Ad hoc* networks combine multiple existing system networks for BAS, meters, etc. into a common data communications network. According to Wikipedia, *ad hoc* is a Latin term meaning "for this purpose." Wikipedia goes on to discuss the mobile internet and Wi-Fi industry for wireless *ad hoc* networks, sometimes called "WANET." The point however is that these are decentralized building networks that are *ad hoc* because they rely on a pre-existing infrastructure of building systems. These systems again include BAS, metering etc., but they also include routers in wired networks or access points in managed (infrastructure) wireless networks. All of these systems are made up of devices that are capturing data about building operations, which are "nodes" on sub-system networks.

The notion of an *ad hoc* network is that a communication infrastructure includes those nodes and systems described above, as well as middleware technology including software and hardware, like routers and gateways, which makes it possible to first communicate to an analytics tool, and second convert (or map) the data to a format that the platform can use to perform analysis. Another way to refer to these systems would be that they enable analytics solutions to access data from large numbers of disparate devices dynamically, through network interoperability. *Ad hoc* networks are predominant throughout the energy and analytics industry, and this is likely to continue to be true. That is because access to the large volumes of data needed by this industry requires access to a large set of networks, made up of large numbers of devices.

## MIDDLEWARE *AD HOC* NETWORK INFRASTRUCTURE

This chapter has provided the framework for a more detailed explanation of what an *ad hoc* network looks like for energy and analytics. As mentioned, there have been numerous approaches to providing solutions by multiple companies. In Jim Sinopoli's article from 2008, mentioned above, he identified several of those companies. This discussion will provide an example based upon one of those companies, Tridium. The reader should not construe this as an endorsement, but rather should note that this company was an early leader in providing technology solutions for *ad hoc* networking. Because of that early leadership they can be used to illustrate actual examples of how many analytics providers create solutions. Again Tridium is not the only technology solution company, the space includes numerous large and small providers from the buildings industry, such as Lynx Spring, as well as from the network and IT industry. Participants in this space range from Cisco Systems and Intel to smaller entrepreneurial businesses like Sky Foundry, Cimetrics, J2 Innovations, Coppertree Analytics and more to come. The reader should recognize that there are nuances to how all of these providers have developed technology solutions, as well as how they approach the market and differentiate from their competitors. In the end however, Tridium is a good representative technology platform to use in discussing how middleware interoperability is achieved for energy and analytics tools. To provide some general background on the company, Tridium is a global software and technology company that created the Niagara Framework®, a software framework that integrates diverse systems and devices—regardless of manufacturer, or communication protocol—into a unified platform that can be easily managed and controlled in real time over the internet using a standard web browser.

This discussion on *ad hoc* networks as created using Tridium products will begin with an overview of a sample application including what data are needed for energy and analytics technology. The discussion will then go on to discuss, in general terms, how the technology might be deployed to achieve this goal. To keep things simple, the application will not be overly complex, and will entail accessing data from energy meters for analysis and from a number of HVAC units for fault detection and diagnostics. The energy data needed begin with a utility billing data history including cost, units of electricity (kWh) and kilowatt demand (kW) for electricity and billing history including cost and units (i.e. MCF) of consumption for natural gas. In addition to energy history, analytics tools will also need 15-minute interval data (typically kW demand) for energy analysis. This will be used to analyze potential benefits from participation in automated demand response and demand limiting programs, as well as the option to sell capacity into electric markets. For equipment optimization and FD&D, the application would provide predictive maintenance (PdM) for 10 large floor-by-floor variable air volume air handling units (AHUs). Repeating some of the content from Chapter 6 to elaborate on this application for these AHUs, PdM would be applied to reduce life cycle energy consumption for the equipment and to positively impact the mission being carried out within the space this equipment serves. As noted previously, countless studies have documented employee productivity improvements in buildings that maintain optimum environments (temperature, humidity, etc.). This application would use PdM for air filter maintenance, by measuring the differential pressure across the AHU filter. The technology would detect filters as they load up with dirt over time (differential pressure increasing) and trigger filter maintenance at the right time. When the filter is replaced, the technology will record the differential pressure drop and hence verify that the replacement was done correctly. By analyzing the pressure drop across the filters over time, it is possible to better establish when, and how often, to change the filter and perhaps even glean information on which filter manufacturer sells a better product. Other analytics algorithms would use the same data on space temperature and humidity, as well as perhaps $CO_2$ and particulates, to compare actual conditions to benchmark metrics established for environmental conditions. Many other factors could be analyzed as well, including light levels and energy metrics at a sub-meter level, etc.

To execute on this application, it is necessary to access data on all of the conditions outlined above. Accessing that data will require interface with a wide range of systems as discussed, so the balance of this section will outline review the analytics network infrastructure as well as the device nodes, and they would be configured. In other words, this is the true meat of the middleware discussion. Again, there are many manufacturers offering solutions in this space, but Tridium will be used as an example only. Interestingly, Tridium and other technology companies have not supported the use of middleware, as a term, to describe the functions that are being carried out with data communications and building systems infrastructure. It is believed that this is because use of the term in the IT industry is too narrow for the functionality that is carried out for applications such BAS, EEM, energy and analytics, etc. Devices like the Tridium JACE™ incorporate much more than self-contained software services for enterprise application. However, as previously mentioned, the energy, buildings and BAS industry have embraced the term middleware to describe the class of products that enables this network infrastructure for BAS and analytics, so it has been used here.

The analytics network infrastructure for the application discussed above could be depicted in Figure 12-5. In fact this is a generic illustration from Tridium Product literature for a product called the JACE-6. Before discussing that product, it is worthwhile to note that Tridium offers a complete portfolio of technologies that extend connectivity, integration and interoperability to millions of devices deployed in the market, but that are capable of being deployed as BAS offerings as well. The product portfolio is delivered in the form of hardware and software, but it is built around the Niagara™ software framework. This software framework is a universal, reusable software platform to develop applications, products and solutions. Software frameworks include support programs, compilers, code libraries, scripting language, an application programming interface (API) and tool sets that bring together all the different components to enable development of a project or solution. This chapter is in no way intended to be a training program for these products, rather it is providing this information to assist the reader in understanding the complexity of these network infrastructures. In the context of energy and analytics, these products are used to resolve the challenges associated with open systems integration and interoperability, by integrating diverse systems and devices, regardless of the manufacturer, or communication protocol, into a unified platform that can be managed and can access data over the internet. The hardware platform that is deployed in the building is a JACE, an acronym for Java Application Control Engine, as shown in Figure 12-5.

A JACE provides the mechanism to achieve interoperability, for access to data, from all systems in a building. By connecting common network protocols such as BACnet™, LonWorks™ and Modbus™, along with many proprietary, or legacy networks, a unified system without seams emerges. Scalability and reliability concerns are avoided with the unique distributed architecture created by a network of JACEs. This is accomplished by interfacing with standard protocols, such as those above, but the company, and members of its distribution channel, also maintain a library of "drivers" for legacy system protocols. At a simple level, the idea is that these devices can be deployed with a driver for almost any legacy protocol for BAS, meters, data loggers, etc. Figure 12-5 shows interface to all of the systems described above for this analytics application: meters with electronic interfaces for energy consumption and interval data, internet access to download billing data from utility portals and BAS systems for equipment data, etc. The reader should also note that, as with BAS, this architecture also shows a Workplace PRO as well.

Previous chapters discussed that BACnet™ and other standards are for data communication, but not for device programming, and that a "workbench" software product is used for that task. This workplace product fulfills a similar role for the architecture depicted, but the intent of this section is not to explain how the devices are programmed to accomplish these tasks. Beyond communication and gateway functionality, JACEs often have inputs and outputs, like any BAS controller, so they can actually be implemented as a direct digital control device complete with programming capability to create algorithms for equipment control. The JACE-6 for example "is part of a family of high performance embedded controllers and includes a fast (524 MHz) processor, 128 MB base memory expandable to 256, and a math coprocessor for better performance. The JACE-6 also comes with the latest NiagaraAX 3.2 software and the Web User Interface, and Niagara Connectivity built in. The product's modular design and flexibility allows for easy installation and distributing connectivity and control more widely across all types of environments." The unit is expandable and features multiple I/O modules (i.e. IO-16 below) and plug communication cards that save engineering, time and installation cost. Like many BAS devices, it comes with optional IO-16 and or IO-34 modules, internal battery backup, two 10/100Mbit Ethernet, RJ-45 ports on the CPU board, a USB port on the CPU board, and two option card slots for communication option cards (Lon, RS-232, RS-485, etc.). In closing, it is important to note that these

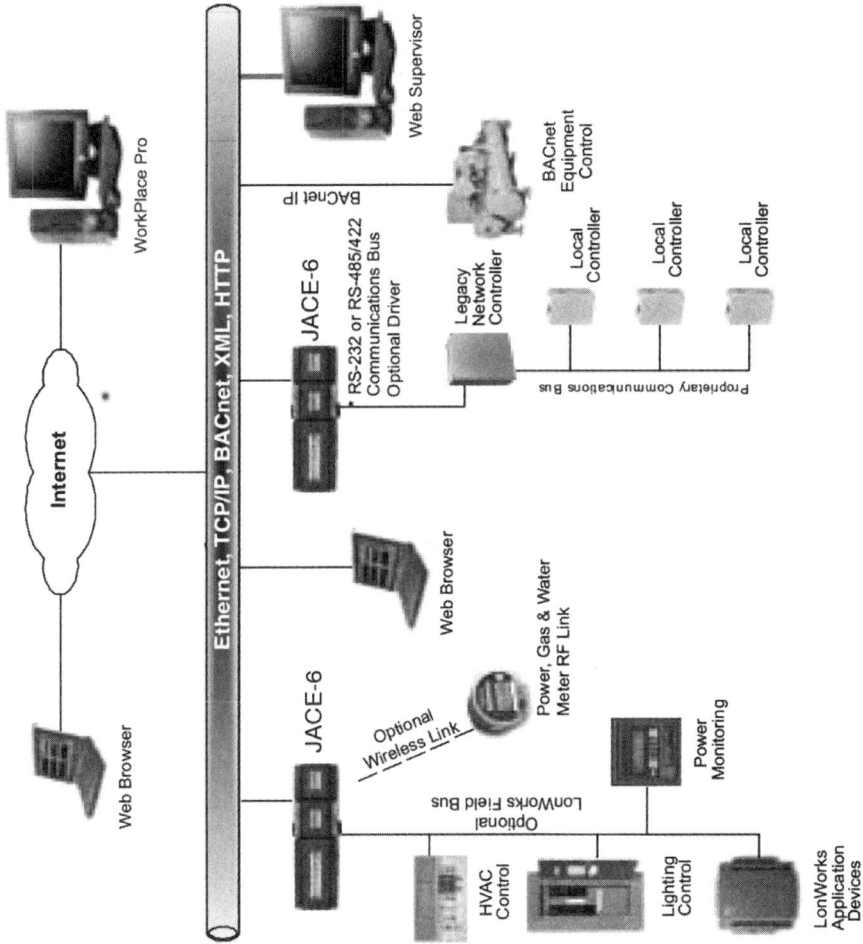

**Figure 12-5**

devices, and those available from a growing number of providers, are highly sophisticated and may be programmed by a capable integrator to make data available for almost any existing building system.

One last topic of particular import for energy and analytics is network resilience. This topic simply emphasizes the importance of a highly sophisticated network design, deployed by a credible integrator. As a side note, that integrator must complete a thorough commissioning which may need to be completed twice, once in heating mode and once in cooling mode, to ensure that this newly implemented system is completely functional in all seasons. Regarding network resilience, the simple challenge with energy and analytics, as with BAS, is that all of these systems typically use corporate networks, as the communication backbone, to simplify network infrastructure and reduce cabling cost. They are also often assigned an IP address so that they can use the corporate LAN to access the internet directly or through a VPN infrastructure. Again as an example, consider Cimetrics, Inc. as a best in class provider to discuss this further. This company was one of the early adopters of BACnet™, and other communications standards, used for energy and buildings solutions. They also offer a product portfolio similar to the JACE. Yet a significant distinction is that the company is also a full-service analytics provider, and has developed best practices to ensure network resilience. The ultimate goal is to achieve very high quality data access, and to enter into agreements that allow the company to perform analytics on equipment *and* the data communications interfaces put in place to access data. As discussed in Chapters 10 and 11, there are a large number of threats to a robust energy and analytics data stream. Quite simply, IT staff and network administrators are concerned solely with the computing infrastructure and business applications. The energy and analytics technology, as well as the BAS, and all of the other special systems in buildings, just do not show up on the IT radar. As a result, IT managers do not notify solution providers when network maintenance is being conducted, or when major changes are made to IT infrastructure, because they either don't think of it or don't even know these systems exist. Many new entrants to the energy and analytics space do not recognize this threat to network resilience, and therefore to their data stream, and they implement one-off middleware solutions, assuming nothing will change. Companies like Cimetrics don't make those mistakes; they leverage IT and network management systems in their design, to ensure that their systems remain viable. Equally important, they structure their design so that the network maintenance staff owns responsibility for devices that will provide data, and they communicate with IT regularly to be sure they plan for upcoming changes. They know that providers cannot tell IT one time that energy and analytics tools are in place and critical to building performance, and expect them to consider these tools every time they upgrade the IT infrastructure. The reader should become schooled in these topics so that they can ensure continuous access to data for analytics therefore protecting the integrity of their systems.

Cyber security is very much in alignment with network resilience, and its importance cannot be stressed enough. This topic was also covered in Chapter 11, but will be revisited here. As with the approach that Cimetrics takes to design resilient networks, energy and analytics staff must be aware of cyber security risks and develop strategies to mitigate those risks. In many respects this is another reason to collaborate with IT staff and ensure that risks are controlled. The intent of this book is not to provide a treatise on cyber security. Readers should know that cyber-crime costs computer users hundreds of billions of dollars annually in systems downtime, time to correct problems, etc., and should expect cyber-attacks to increase. Tools and techniques exist in the marketplace to address both the volume and complexity of cyber-attacks, so that enterprises can stay viable in the face of evolving threats. Combining energy and analytics with security technologies yields a stronger defense posture. Security analytics can provide high-speed, automated analysis to bring network activity into clear focus to detect and stop threats, and shorten the time to remediation when attacks occur. The reader is encouraged to become knowledgeable in this area and to develop strategies that protect access to data. Such strategies will also reduce down time and protect the building mission as well.

The last topic in this chapter is web services. In fact, web services have been mentioned numerous times so far in this book and this chapter. It is a term that industry professionals have become very familiar with, and it is an approach that is integral to many energy and analytics solutions. Given the importance of web services to this topic, it is worthwhile to provide some basic definitions. This is also important because coming chapters will talk about system integration and analytics tools, so may be of value to provide some content on how this technology may be architected.

So what are web services? Vangie Beal, managing editor of Webopedia.com, published this content on the term web services: "Web services describes a standardized way of integrating web-based applications using the XML (Extensible Markup Language), SOAP (Simple Object Access Protocol), WSDL (web Services Description Language) and UDDI (Universal Description, Discovery and Integration) open standards over an internet protocol backbone. It is common for web-based applications to use XML, which is designed especially for web documents to tag the data, SOAP is used to transfer the data, WSDL is used for describing the services available and UDDI is used for listing what services are available. Used primarily as a means for businesses to communicate with each other and with clients, web services allow organizations to communicate data without intimate knowledge of each other's IT systems behind the firewall. Unlike traditional client/server models, such as a web server/web page system, web services do not provide the user with a GUI. Web services instead share business logic, data and processes through a programmatic interface across a network. The applications interface, not the users. Developers can then add the web service to a GUI (such as a web page or an executable program) to offer specific functionality to users. Web services allow different applications from different sources to communicate with each other without time-consuming custom coding, and because all communication is in XML, web services are not tied to any one operating system or programming language."

Another definition for web services comes from Wikipedia, the free encyclopedia, which says that a web service is a method of communication between two electronic devices over a network. A web service is a software function provided at a network address over the web with the service *always on* as in the concept of utility computing. The World Wide Web Consortium (W3C) defines a web service generally as: a software system designed to support interoperable machine-to-machine interaction over a network. The W3C Web Services Architecture Working Group defined a web services architecture, requiring a specific implementation of a "web service." In this a web service has an interface described in a format that can be machine-processed (specifically WSDL). Other systems interact with the web service in a manner prescribed by its description using SOAP (simple object access protocol) messages, typically conveyed using HTTP with an XML serialization in conjunction with other web-related standards.

Clearly there is a great deal of content regarding web services. The author's goal however is to provide very basic information that will be useful to readers who plan to specify, buy and operate energy and analytics systems. Therefore the goal is not to understand how to develop such a system, but to understand the underlying technology that is being purchased, and how that knowledge may be used in the design/purchase process to acquire the right solution.

Given the introduction to web services, it is most important to know how this technology is used with energy and analytics. The next two chapters will consider integration and analytic tools and product configurations, so this section will not go into great detail on those topics, rather it will focus how the tools are offered. This is a marketing topic related to the product offer and how it is priced. A very popular product approach today, across multiple technology industries, is "Software as a Service" (SaaS). SaaS is exactly what is being offered with most energy and analytics tools, and it is being offered using web services. The software in this case, is the energy and analytics tool, and the SaaS approach defines how the software is structured, packaged and offered to customers. In terms of structure, SaaS can be very cost effective because the computing infrastructure made up of servers, processing power, storage (computer memory) and application software itself resides off-site. Another way to say this is that the software resides "in the cloud," and the service entails supporting all of the computing infrastructure and making it available through an internet-based web service. The energy and analytics tools provider therefore offers access to the software and computing infrastructure for a recurring fee. Of course this is a two-edged sword, these benefits are one edge of the sword, and the other edge is that these products are not "shrink wrapped" for one-time purchase and ongoing use, the SaaS model is to sell "use" of the software for that recurring fee. So the SaaS model for energy and analytics tools typically has a service, or SaaS, contract with an ongoing monthly, quarterly, etc. charge. That charge is use of the tool and computing infrastructure described above, and also includes ongoing upgrades, etc. to the tool and the infrastructure. As noted in the definitions of web services above, it is common for products to use XML and SOAP, etc.; however, for buildings applications targeting dashboard and analytics, XML and SOAP may be more sophisticated that requirements dictate. As a result, some simpler solutions have become available that are easier to develop products around and are easier to support. At this writing there are a few popular options to simplify solutions and two will be discussed: Representational State Transfer (REST) and JavaScript Object Notation (JSON).

**Representational state transfer (REST)** is a software architecture style consisting of guidelines and best practices for creating scalable web services. REST is a coordinated set of constraints applied to the design of components in a distributed system, which can use text, graphic and video to provide a more user-friendly and maintainable architecture. REST has gained widespread acceptance across the web as a simpler alternative to SOAP and WSDL-based web services. Systems that deploy REST typically, but not always, communicate over the hypertext transfer protocol with the same HTTP verbs (GET, POST, PUT, DELETE, etc.) used by web browsers to retrieve web pages and send data to remote servers. The REST architectural style was developed by W3C Technical Architecture Group (TAG). The World Wide Web represents the largest implementation of a system conforming to the REST architectural style. May energy and analytics systems, such as SkyFoundry are integrating "Native" REST.

The web services technologies discussed here are in some cases standards, and in others software building blocks to develop product offerings. It is also important to note that integrators are often faced with the challenge of creating "glue" to enable interface between existing systems and a web services offering. To accomplish this functionality, an API (application programming interface) is often used. One such product is the Cimetrics B6080, which is a BACnet interface to web services gateway/firewall. This product allows integration of BACnet/IP (internet protocol) devices in enterprise application systems using REST architecture. The device uses Cimetrics BACrest API (application program interface). It is a specification that allows applications to read and write data to device-centric networks using URL (uniform resource locator)-like commands. According to Cimetrics, "the meaning of the word BACrest is building automation control relational state transfer. The "REST" part of this name is a well-known IT terminology that describes the architecture used to create the internet "URL system." URL is a synonym for Web address. They have created a version of "REST" that applies to the needs of device-centric networks—the simplest yet most powerful data exchange methodology. This provides a more capable and secure interface for enterprise applications allowing access to more of the data found in building automation systems. Using this API approach, it is possible to integrate BAS with other enterprise computing applications using SOAP/XML technologies. This allows platforms such as Microsoft.Net or Java to access real time data from a building automation system. Since two isolated Ethernet ports are used, this device also acts as a firewall, meaning that no unintended communications or control is possible between the two networks. This can be a valuable feature to develop strategies that protect against cyber security threats.

Of particular interest for energy and analytics is that The Project Haystack data modeling standard has provided a documented representational state transfer, application programming interface (REST API) to define a simple mechanism to exchange "tagged" data over web services. REST servers are programmed to implement a set of ops or operations. An operation is a URI (uniform resource identifier) that receives a request and returns a response. The URI is an HTTP URI, which is used at a base address. Operations are then mapped as path names under that address. Standard operations are defined to query databases, set up subscriptions, or read/write histories of time-series data. Operations are pluggable so vendors can enhance open REST interfaces with customized, value-added functionality for their own business purposes.

**JSON** is another approach used by a number of technology companies in the energy and buildings space. JSON "is a lightweight data-interchange format," according to the JavaScript Object Notation Website. It is easy for humans to read and write. It is easy for machines to parse and generate. It is based on a subset of the JavaScript programming language, Standard ECMA-262 3rd Edition, December 1999. JSON is a text format that is completely language independent but uses conventions that are familiar to programmers of the C-family of languages, including C, C++, C#, Java, JavaScript, Perl, Python, and many others. These properties make JSON an ideal data-interchange language.

JSON is built on two structures:

- A collection of name/value pairs. In various languages, this is realized as an *object*, record, struct, dictionary, hash table, keyed list, or associative array.

- An ordered list of values. In most languages, this is realized as an *array*, vector, list, or sequence.

These are universal data structures. Virtually all modern programming languages support them in one form or another. It makes sense that a data format that is interchangeable with programming languages also be based on these structures."

Again, the technical details of how products might be developed and operate using JSON is beyond the scope of this book. However, it is useful for the reader to understand that tools like this exist to simplify the structure of web services offerings for energy and analytics. The result will be faster time to market for new products and faster upgrade and enhancement for existing ones, in the end this all benefits the user and operator.

In closing, this chapter has covered a diverse set of topics related to the infrastructure of energy and analytics technology. The reader should note that this "middleware" segment is in a tremendous state of flux. As discussed, these products were first introduced to the market as independent components that required separate interface, programming and installation. An analogy for consideration in coming to grips with the complexity of implementing this technology might be to local area networks (LANs). LANs are pervasive for our IT applications, as well as for internet access and web services. Yet when considering those LANs, the user/manager must understand that there are independent skills, and in many cases independent service providers, who must be hired to install and maintain separate system components. For example:

- The servers, PCs, etc. have one set of requirements for deployment,

- The LANs themselves including all of the network software, routing, switching and cabling infrastructure, while application software have another set of requirements and

- Internet service providers (ISP) have yet other requirements.

One company is typically hired to provide and implement the computer technology, a second set of companies will be hired to provide software (i.e. office suites, accounting systems, etc.) and a third company will be hired for the LAN. It is understood that companies like CISCO, 3Com, etc. have very unique skills and are certified independently from the other providers, but typiaclly cannot provide an entire end-to-end computing system complete with applications. This is true, just as an accounting software company for example would support its application, but would not be called to install a LAN. Of course companies differentiate in many ways, but the bottom line is that users understand that these products require separate content expertise.

In light of the LAN example, now consider building technologies particularly for energy and analytics. Over the years it has been common in the building space to expect one provider, often called an integrator, to combine content expertise across all applications (BAS, dashboards, energy management, analytics, etc.), as well as to be the LAN, ISP and web services provider. In many cases, each of these products and applications is manufactured by separate companies, but the building owner/manager typically uses one system integrator as the technology content expert for complete implementations. As time has progressed, the idea of bolt-on middleware components that are applied, independent of the building systems, has begun to morph, but the role of the integrator has remained constant. Many IT companies have announced plans to dominate this industry. Interestingly however, these companies usually end up coming back to the same system integrator channel to implement their solutions, because this is the only group that has the content expertise required to deploy the products properly in a building. This may or may not change over time, but one thing is changing, and that is how energy and analytics products are deploying middleware. Some companies are combining this technology into their products to simply expand the functionality, while others offer independent products for each function. The important take-away from this chapter is not how middleware is configured in particular products, or how it works today, but that middleware is, and will likely remain, a critical component for energy and analytics technology. As discussed in the beginning of this chapter, for energy and analytics middleware is essential to controlling the cost of systems implementation. For the building owner it just makes sense to utilize sensors and instrumentation, which reside in systems that have already been purchased, as sources of data for energy and analytics. This approach reduces the overall cost of acquisition for energy and analytics, while also speeding deployment. For that reason middleware will continue to be important to this topic and the reader should understand it well enough to ask informed questions about how data will be acquired and whether they can reasonably expect that data acquisition to be reliable.

# Chapter 13

# System Integration

The previous three chapters treated technologies related to BAS, data communications/networking and middleware in some detail. This chapter will not repeat that content, but will expand on that discussion by introducing the concept of "system integration," or simply integration. Integration implies a broader "system-wide" context, and is the practice of combining systems to achieve some level of "connectivity" or "interoperability." These terms have been used in very similar ways, almost interchangeably, in the past, but the term "interoperability" has become more universally used to describe interaction between smart systems. In the IT world there are levels of interoperability, but the focus here is on energy and analytics with systems in buildings. Therefore this chapter will focus on two levels of interoperability: data access and control integration.

Under the heading of energy and analytics, the primary concern is data access from many sources, as has been discussed. Throughout the last three chapters there have been many references to integration and system interaction. The purpose of this discussion is to address the primary issue: installing instrumentation and networks to capture data is cost prohibitive and unnecessary when exiting systems can be mined for much of the necessary data. Mining those systems is what will be referred to as "integration." Clearly there will be cases when specific data are required that cannot come from another system. In that case a decision can be made whether to add points to existing systems and access data in the same way, or to add new instrumentation. This chapter's sole purpose is to emphasize that much of the data required for energy and analytics technology can come from integration to BAS, as well as a larger ecosystem of building technologies. Other potential sources for energy and analytics suites can also include meters, CMMS, HVAC security and a host of systems that operate independently of the traditional BAS, but are essential to both, capturing data for analysis. The same systems may also be used for control integration, which is executing changes to facility operations that are recommended by the analysis.

System integration is a term that has been used extensively with BAS. It actually pre-dates the introduction of BACnet™, as a reference to this second topic "control integration."

Control integration is not the focus of this chapter, or this book, but some time must be devoted to this topic because over time energy and analytics technology can be expected to result in execution of real-time corrections, as well as report recommendations. The process today is that the analytics tool executes algorithms to analyze performance and identify anomalies; the next step is to report and ideally prioritize issues, based upon customer criteria, and make recommendations. In reality, many of the issues identified will require more detailed human interaction to identify root causes and resolve problems, however some corrections could happen automatically. For example, it could be possible to clear BAS system overrides automatically, and some control variables could be modified automatically based on learning how HVAC performs on a day-to-day, month-to-month basis. Of course the devil is in the details with all of these types of sequences. However, BAS algorithms like optimal start stop have been doing exactly the same thing for decades—learning from performance and modifying start times based upon conditions.

The new electricity market creates even more options to execute pre-approved shed sequences on kW demand status, and to fine tune those responses over time. What is really exciting about energy and analytics is that the technology can monitor many more variables than ever before. An example of such a sequence involves the way that building occupancy varies during the week. Building owners, especially those in large cities with difficult commuting, have been adapting to the growing participation of workers in telecommuting, which varies throughout the week and the year. It is very common for office buildings to be very lightly loaded on Friday afternoons, Monday mornings, etc. A great example of how applications like these are being expanded to include integration with a wider range of systems was pointed out in Jim Sinopoli's article, "Predictions for 2015" in February of that year (www. automatedbuildings.com). The article discussed Bluetooth Beaconing, a new technology for "indoor positioning, which can be used to get accurate data on building occupancy at any time based on smart phones. Beacons use Bluetooth Low Energy (BLE), which is built into most of the latest smart phones, and this could revolutionize how occupancy data are determined within a building, at a very low cost. This information opens a host of opportunities for building systems, such as HVAC, but how about cycling elevators because there is no need for four or more to be operating if the building is 50% occupied? In the past, these algorithms required occupancy data from security or lighting controls, as well as trending data on elevator operations. With BLE integration this could be done automatically without waiting for secondary analysis of the data from other systems. There are a wealth of new control integration opportunities, and these can be expected to increase exponentially as web services and integration becomes possible with many previously "siloed" systems.

## BUILDING AUTOMATION SYSTEMS (BAS) AND CONTROL

In prior chapters, BAS technology was described in detail, but this chapter will begin with a simple prediction: The future of the BAS industry will not be determined by a small group of large control manufacturers. The ever expanding content knowledge of system integrators coupled with new technology will accelerate movement toward this new business reality. The BAS industry now includes companies offering many types of software and hardware, including middleware. Equally important, the role and functionality that is introduced by web services expands the power of these systems. Industry leaders agree that automation and information technology (IT) have converged, and yet neither the conventional BAS players, nor the IT companies entering this space, are obvious winners. At no time since the market explosion caused the oil embargoes of 1970s have BASs, then called energy management systems (EMS), seen such a diverse and robust marketplace. The previous hardware and software barriers to entry for BAS players have been removed due to the proliferation of microprocessor technology and system standards, while web services development tools have also made it possible to launch offerings with much lower up-front cost.

A very robust ecosystem of technology companies is participating in this market, as a result of the market changes described, and that includes energy and analytics. There are pros and cons to this market environment. Obviously it is great to have many choices, but it is also harder for the companies to differentiate and harder for the owner/manager to choose wisely. The clear outcome, however, is that new advances in building systems technology are less likely to come from the old brand name companies. In fact, advances are more likely to come from startups and small players, and this has fueled a merger and acquisition frenzy in the market as well. This is evidenced by Honeywell's acquisition of Tridium and Akuacom, Johnson Controls acquisition of Grid Logix and Energy Connect, and Schneider Electric's acquisition of TAC and Power Measurement. Therefore it is important to consider how industry participants look at the state of the BAS business. Traditionally the major industry participants were control manufacturers that developed BAS products, which have evolved from simple mechanical and electrical devices to fully networked suites of smart systems. These large companies were multinational and multidisciplinary in nature, but their vertical integration encompassed more that the BAS. Vertical integration in this case means controlling all of the raw materials, supply chains and product offerings that are necessary to offer BAS, but now these companies are also acquiring the equipment that is controlled as well. The major players are actively aligning their BAS offerings by acquiring heating, ventilation and air conditioning (HVAC) companies, just as the dominant HVAC players have acquired BAS companies since the 1980s. This may be somewhat interesting, but the reader may ask, "Why do I care?" The reason to care is that many managers and owners of buildings, particularly those that own multiple buildings in multiple locations, are faced with a dilemma. It is becoming harder and harder to rely on any one company to provide end-to-end service, and this means that managers must become more tech-savvy and better able to understand how to take control of their own BAS, energy and analytics destiny. Understanding integration is the most important first step that any manager can take in that progression. Readers who have reached this point in this book are well on their way to developing that understanding. Of course, this knowledge must be tempered with real-world experience, and requires ongoing effort for those who want to remain current.

ARCHITECTURE AND INTEGRATION

The logical place to begin understanding integration and its role with energy and analytics is to revisit the topic of architecture. From the previous three chapters the reader understands that architecture includes hardware and software, but equally important relies on an interwoven series of standards for data communications that must be compatible and interoperable. If the data communication protocols being deployed within the architecture are not interoperable, it will be necessary to abandon any data that could be accessed or to implement a solution, such as middleware technology. Therefore step one in the design of any energy and analytics solution will be to identify the business processes that require analysis, to identify the data that are needed to conduct that analysis and to determine if the data are available and resident in an existing system. If the data reside in an existing system, integration will be necessary. This is the beginning of developing the energy and analytics architecture. For each system component, web service or other data source, an element of architecture must be created and a strategy must developed to access that data. Data quality is important because access to the data must be reliable and continuous. With heightened concern about cyber security, this architecture development must also consider whether firewalls are in place, and if it is possible to work with IT and network administration staff to devise a method for interfacing to the systems that hold the data.

As stated, the term used to describe this overarching process of overcoming obstacles to interface with systems is "system integration." Definition of this term has evolved over time, and what we mean by system integration includes what we call building automation. For example, integrated systems combine HVAC control with fire/life safety and security systems. Actual integration of control sequences ranges from extremely limited, to complex approaches for smoke evacuation and after-hours access to selected areas of a building. Over the years, system integrators have evolved, and the scope of their services includes all building systems, as well as web services, energy and analytics, visualization/dashboards and a host of customer-specific services. At the simplest level these integrators develop business content expertise to help the customer identify problems that can be solved through combined system functionality. As mentioned above, that may be accomplished through data access or through a combination of data access, data analysis and integrated algorithms or "control integration." These services require that integrators be technology experts with all building systems, energy experts, and that they also be network integrators. This breadth of skill is what makes it possible to determine how an architecture must be designed and implemented to achieve optimum performance from energy and analytics.

The expanding technologies that comprise integration require industry participants to revisit training needs. Knowledge necessary to be successful in building automation starts with an understanding of BAS, web services and much more. That sentence covers a tremendous amount of ground and, in a world that demands specialization, there are two or three areas to which a person could devote an entire career. In the real world however, the industry demands more, so after mastering BAS, the system integrator must also understand data communication, networking and web services.

System integration also presents challenges for the construction procurement process on numerous levels. The industry is demanding integration throughout the enterprise—from building automation, fire alarm, access and video surveillance for security, to energy management for the enterprise, and metering and maintenance management. System integration goes beyond independent building systems such as those listed above, because it requires commonality with building application infrastructure including hardwired and wireless LANs, central databases and even network-client software, among others. The dramatic industry effort to develop standards for networking and communication has resulted in a host of BACnet™ and LON™ compatible products, but with web services and the smart grid that world is expanding dramatically. As previously discussed, the smart grid interoperability panel (SGIP) is identifying and codifying an ever-expanding number of standards that are at the integrators' disposal.

Training is, of course, needed to fully access all of the possibilities, and sources for this training are becoming available. Many technical colleges and trades associations offer seminars and short courses covering specific topics in system integration and the related data communications and web services topics that have been discussed here. Www. automatedbuildings.com launched online courses in the past, but the focus now is on up-to-date content from experts, in the form of articles and interviews. Training, however, is an important focus for trade associations such as the Association of Energy Engineers (AEE) which offers a host of face-to-face and online seminar programs. AEE face-to-face seminars are held around the country and are often co-located with their popular conferences. The online seminars are offered in both synchronous (everyone attends simultaneously through an online meeting web service) or asynchronous training environments (which allow students to learn at their own pace), while accessing real-time information from other participants via virtual seminar meetings. Training offerings should be evaluated based upon what content area the participant needs to learn about. An important consideration is that integrated systems, as discussed in relation to energy and analytics, include systems that encompass a wide range to technology elements. Again readers should look for training options that highlight their specific needs.

## ENERGY MANAGEMENT, SUSTAINABILITY AND ANALYTICS

System integration that has been coupled with energy and analytics was traditionally call enterprise energy management. At its most elementary level this is primarily a technology that allows for monitoring of utility meters and energy data to develop baselines and benchmarks. Again this information can be very important for a host of functions. Ever-expanding functionality for these systems has broadened however, allowing energy and analytics to be part of full-scale integration of BAS and other building systems, as well as utility procurement, demand response, and participation in other utility programs plus effective use of onsite generation. This is clearly the future of energy management and expands the definition of system integration to its logical conclusion. At the expense of repetition, it demands that system integrators be network integrators as well. Taking this topic to its logical conclusion, one of the hottest topics today, planet-wide is sustainability. The tie between energy and climate issues is unquestioned, but the opportunity for energy and analytics to both measure and improve sustainability efforts is also significant. There is a compelling and dramatic interrelationship between the supply and demand sides of energy. This relationship has been discussed a great deal with regard to electricity in this book, but it is equally compelling as it relates to oil and natural gas. At this writing, the price of oil is just above $50 per barrel after nearly 5 years at or above $100 per barrel, and the market is highly volatile. Prices are being driven by market strategy, again on behalf of middle-eastern oil producing nations, but also by unprecedented access to low-cost oil and natural gas in the United States. For purposes of this discussion, the focus stays on the building and its owner or manager, and how energy and analytics can optimize energy performance within the building. With efficiency and operational actions achieving the best possible performance, the manager will be able to fully leverage options on the supply side more effectively. Those supply-side options may be electricity and/or natural gas procurement, participation in DR or energy markets, implementing DG or any number of other options.

Clearly an unparalleled opportunity exists to leverage BAS and integration for intelligent buildings that can fully leverage the energy opportunity. The utility industry started down this path with early smart grid strategies, but then became sidetracked building out the advanced metering infrastructure (AMI). This was a multi-billion initiative that utilities throughout North America embraced, resulting in web-enabled meters that were intended to transform the electricity business. The idea of real-time pricing, and such things as "prices to devices" to enable real-time pricing, have been deployed in the more advanced electricity markets. These markets are almost universally in states that have deregulated electricity markets, which enable energy users to make buying decisions to purchase from a variety of sellers. These buying decisions may also be extended to time-based procurement that analyzes the actual current value of the energy commodity, i.e. electricity. By the way, many of the AMI meters were deployed with very little functionality. They are like iPhones™ with only two apps, 1) the meter can be read remotely and 2) the power can be turned off if the bill is not paid. So here is the gotcha: it will take decades for expanded AMI apps, and truly price-based buying options, to become a reality for all electricity users, but in the interim, BAS integration and web services can provide the same and even greater functionality in the buildings sector. This has led to the development of system integration as a business model and to companies known as integrators to develop value propositions that sidestep the utilities and drive results. Early adopters were participating in demand response a decade ago, and this will continue to adapt and grow. Demand response is a multi-billion dollar market, and savvy system integrators have redefined themselves to deliver smart energy building solutions and to set the standard for a global solution to climate change. Energy and analytics are integral to these solutions. System integrators engaged in this higher-level thinking are considering the intersection of buildings, energy and environment to illustrate a point: buildings are at the heart of our energy and climate issues. Recent data from the Department of Energy indicate that electricity generation is responsible for 39% of greenhouse gas emissions, and the American Institute of Architects credits buildings alone with responsibility for 48% of greenhouse gas emissions. There is some overlap between these numbers since buildings are large electricity users, and the burning of fuel to produce that electricity is what causes part of the greenhouse gas emissions contributed by the building sector. There has also been solid research done to validate the fact that these emissions increase at times of peak electric demand. This is because utilities must fire their least efficient units to meet peak electric demand, and of course those higher polluting coal-fired power plants are at full capacity. This is really just another way to embrace the whole idea of green buildings, but more importantly to emphasize that HVAC, automation and energy efficiency are highly important parts of the equation. System integrators can't tell a story about how their work addresses building siting, urban redevelopment, bike racks or recycled carpet, but they can tell a story about energy and analytics, efficiency, and demand response.

The intent of this chapter has been to refocus on system integration as an enabling skill set and a set of technologies essential to deploying energy and analytics. The initial steps remain critical: 1) develop requirements, and 2) evaluate the architecture that exists, along with the system integration that is necessary to modify that architecture. System integrators are instrumental in determining how to transform systems from those that operate in independent silos of information, to those that can interoperate, and how to allow managers to access the data that are needed. The next step will be to determine whether it is also possible to develop integrated control functionality that is responsive to results from energy and analytics, as well as to push that data upstream to decision makers for prioritization and planning.

# Chapter 14

# Project Haystack Data Standards

This chapter focuses on an exciting topic for energy and analytics, Project Haystack and the event it sponsors, Haystack Connect. The formation and activities of this effort are recent and in a state of rapid growth. Project Haystack was designated with 501C3 private nonprofit status less than a year before this book was published, and had held only one national Haystack Connect event. Given the emerging nature of Project Haystack, this chapter will offer on an overarching introduction to the subject including discussion of the drivers that resulted in its creation, following a brief but more specific discussion of its work. As Project Haystack continues to grow and evolve, the expectation is that its list of accomplishments will expand dramatically, and readers should follow their work. John Petze, who wrote the forward to this book and contributed the Introduction to Analytics chapter, is a partner with Skyfoundry, Inc., and is one of the founders of Project Haystack. Mr. Petze has written extensively on this topic and speaks regularly at conferences. Readers are encouraged to attend one of his presentations if possible and to read his articles on www.automatedbuildings.com and in other publications, as well as to visit www.haystackconnect.org to learn more.

## CHALLENGES WITH DATA

The most important point for energy and analytics is that Project Haystack will likely transform every aspect of the work that has been described in this book, and serve to remove a significant obstacle to having this technology become pervasive. This implementation obstacle is caused by the lack of data standards to make it easy to access, understand and use data. This standards void drives higher implementation cost and slows the speed of adoption for energy and analytics technology.

237

This book has highlighted two important facts about energy and analytics: 1) tremendous value that can be achieved by leveraging data resident in existing systems, and 2) leveraging these data is critical, because of the high first cost to install new sensors, etc. This high cost of new sensors and instrumentation to create a data stream has been an obstacle to broader analytics deployment. Project Haystack came into being because a group of industry visionaries recognized that a problem existed with accessing data and understood how to solve the problem. Even more unusual, these visionaries were willing to develop the solution in a collaborative community, rather than to develop it as intellectual property for one company. This chapter introduces Project Haystack but it cannot provide deep technical coverage of all their work; however, this author believes it will have a transformational impact on energy and analytics.

This content may be helpful to product developers, but it is intended for system designers, implementation integrators, building operators, energy managers and consulting engineers specifying system requirements. This group may not need to be versed with the specifications and data model that has been created by Project Haystack, but they must be aware of its work and the concept of making data self-describing. A good comparison to this effort might be data communication standards, such as BACnet™, that have become pervasive. For example, most managers are not fluent with BACnet™, at the protocol level, but they understand its importance to the industry, and because they made it a purchasing requirement it has seen such widespread adoption. Another example might be the "Intel Inside™" concept, because it defines a core requirement that consumers have assigned a value to, and because it has shaped buying decisions. This author believes that Project Haystack's work is creating a standardized methodology that can be used across virtually all types of systems and data sources, which will become core requirements for a wide range of products from BAS to smart sensors to analytic tools and other building technologies. A second and equally important result of their work will be the education of an industry, perhaps multiple industries, to the importance of standards for what is known as "data semantics."

To be more specific, Project Haystack is focusing on an industry data challenge that has to do with conveying essential meaning about energy and building data. This concept is known as meta data and semantic modeling in the computer science realm—potentially daunting terms, but ones that address a very simple need, the need to include descriptive information with sensor and machine data so that external applications understand the meaning of the data. For example, is a sensor value a temperature for an occupied room or a return air duct? The Project Haystack community, which includes people from a wide range of domains has developed a flexible, extensible methodology to define and attach semantic data to all types of energy and building data.

Addressing this data modeling challenge has been difficult for a number of reasons. Most significant is that virtually none of the standard protocols that emerged in the buildings industry in the last 20 years addressed the meta data challenge. Because there has been no standard approach, implementation costs for applications that consume data from different systems have been inordinately high. This is because the data must be converted into a format that can be understood by each application such as analytics. This may sound very similar to the protocol topic that was discussed in an earlier chapter, because on the surface it looks like a formatting issue. However in this case, the focus is on data characteristics and structure that allow a system to understand the meaning of information, from its unit of measure, to the equipment system to which it is attached.

Before Project Haystack, it was necessary to convert data from building systems into a format that could be understood by analytics software, and that conversion process is a "one-off manual task" where human operators add essential descriptors to the data items. The one-off task is labor-intensive, and this is what adds more cost to implementations, but it can also create a threat to long-term data resiliency. The threat comes into play because changes to any of the systems that were part of this one-off approach can, and do, occur for any number of reasons, and that can disrupt the data flow. The Project Haystack website points out that operational data, from building systems, etc., has poor semantic modeling and requires a manual, labor-intensive process to "map" the data before value creation can begin. This mapping activity is what technologists call the "one-off" data conversion approach. Project Haystack's work focuses on pragmatic use of naming conventions and taxonomies that can make it more cost effective to analyze, visualize, and derive value from operational data. There will be further coverage of this topic in the chapter on analytics tools. Suffice it to say the data semantics challenge has been a huge issue for energy and analytics tools and Project Haystack could not have been formed at a more opportune time.

To bring the work being done by Project Haystack into focus, the project Vision is to create a standardized methodology for describing data that will make it easier and more cost effective to analyze, visualize and derive value from operational data. This of course is critical for energy and analytics. The Project Haystack website provides more detail by outlining that it "is an open source initiative to streamline working with data from the Internet of Things. The project standardizes semantic data models and web services with the goal of making it easier to unlock value from the vast quantity of data being generated by the smart devices that permeate homes, buildings, factories, and cities." These smart devices are part of applications that "include automation, control, energy, HVAC, lighting, and other environmental systems."

The Project Haystack website is very informative and provides an online forum that the community can use to collaborate and share information. Another good resource for the reader, if this topic is new to them, is an 8-minute YouTube video created by John Petze that may be viewed at: http://youtu.be/5C6GwLbYqTw. Again, the video points out that the challenge for energy and analytics is that most data contained in BAS, and other systems, suffer from poor "semantic modeling," which means that the information that describes the meaning of the data is either not easy to decipher, or in many cases is not available at all. Again as with data communication and networking, standards are essential to making data easy to interpret, and require time-consuming and costly data mapping so that it can understood by applications.

## PROJECT HAYSTACK APPROACH AND ACTIVITIES

The Project Haystack community has outlined some lofty goals, and is already hard at work achieving results. Among their first steps was to up a Project Haystack website at www.project-haystack.org populated with very useful information to educate the industry, and a discussion forum that allows constituents from across the world and different industries to work together to create consensus-driven solutions. Through that effort and a lot of work behind the scenes, a first major accomplishment was for this community to conceptualize the data problem and frame the discussion to identify activities that are required to create a solution. This was done in alignment with the Project Haystack mission, which is to "...define a methodology and common vocabulary so that models of building systems and smart devices can be interpreted automatically by a variety of software and web-based applications." To state this another way, this approach provides a simple, standardized way to represent pieces of information—data. However it is more than just representing numeric values or alpha text; the software applications that use these data must be able to understand the context for the data and their characteristics. To make this possible, Project Haystack created a model, which is a method to organize data in a meaningful way.

The idea is to begin at a high level to define characteristics that explain the context for the data. The Haystack model does this by using "tags" to identify groups of data. Groups of data associated with a building or piece of equipment (i.e. air handler) use an "entity" tag, which identifies and helps to create context. There are also tags to describe the entity. Drilling down into more specific content, there are tags to identify a variety of types of information and to identify characteristics of that information (what type of number is it, is it a time of day, etc.). Simply put, tags define facts about each data item—is it a temperature sensor? Is it a room sensor or a return air sensor? Which air handler is it associated with? What is the schedule associated with that air handler and the spaces it serves? This process standardizes data so that applications can automatically interpret data, without manual intervention, allowing for the actual value add of analytics to be realized cost effectively.

The project also provided a more technical explanation of their goals for energy and building data. "The Project Haystack data modeling standard for Buildings and Equipment systems shall use a simple meta-model." As defined by Wikipedia, "meta-modeling is the analysis, construction and development of the frames, rules, constraints, models and theories applicable and useful for modeling predefined class of problems." In this case the problem is that data from building systems do not follow consistent rules and are not easily used by applications, hence the need for mapping. As noted, the Project Haystack meta-model is based on the broadly accepted concept of "tags." To provide more detail on this concept, tags are name/value pairs, associated with entities like AHUs, electric meters, etc. Tags are simple and dynamic, add structure, and provide the flexibility needed to establish standardized models of diverse systems and equipment. Tags are a modeling technique that allows easy customization of data models on a per-task, per-project or per-equipment basis, while retaining the ability to be interpreted by external applications using a standard, defined methodology and vocabulary. Using Haystack, external web-based applications are able to receive data that include essential meta data (tags) to describe the meaning of the data. This approach can be used for graphics, analytics, maintenance management and other applications, and it enables automatic interpretation of the data by software applications. This Project Haystack effort is a gift to the energy and analytics world, and it could only have been done with the expertise of this highly competent community of technologists and entrepreneurs.

The project also developed a guide specification: http://www. skyfoundry.com/file/50/Guide-Specification-for-Data-Modeling-Standard.docx. This is a highly educational document entitled "Guide Specification for Data Modeling of Building Systems and Equipment Based on Project Haystack Open Source Data Modeling Standard." In the end, to get full-scale market adoption of data modeling standards, this effort is a combination of push-pull strategies. Manufacturers and software/applications developers, such as SkyFoundry, that were early adopters of the data modeling standard are the industry leaders. The broader the adoption of the standard by others, the faster cost and implementation obstacles to analytics will be removed. At the same time, there has been a growing market pull from managers that want access applications for energy and analytics as well as maintenance management and many other applications mentioned in this book. The data model and guide specification offer significant value to all of the market participants.

Project Haystack has also been engaged in development of many technology and business enablers. These enablers take the form of protocols and tool kits to accelerate the development of new products, and the upgrade of existing ones to be compatible with the Haystack Open Source Data Model. These efforts are targeting existing *de facto* standards and widely deployed technology such as the Niagara Framework™ from Tridium, Inc., and JAVA, a high-level object-oriented programming language developed by Sun Microsystems that has been widely used for product development. Taking this approach, it is possible to reduce the cost barriers to participating in this exciting business, and to make it easier for building operators to get the full benefits of energy and analytics.

The last topic under this heading is Haystack Connect. At this writing, final preparations are underway for the 2015 event, and it has become the must-attend event for those with a passion to unlock the power of energy and building data. It is billed as "the place where the community of automation and IoT professionals come together to learn and share the latest techniques for connecting systems and using data to advance the efficiency of buildings, equipment systems and processes." To say it is a conference would be an understatement, it is the educational opportunity of the decade. After the inaugural event in 2014, Ken Sinclair wrote a review of the event saying that for data "the most significant advances are being driven not by a single company, but rather by a community of companies creating open, best-of-breed technologies that work together through a range of open protocols and software interfaces, including the rapidly growing "Haystack" standard. These new technologies reach the market through a growing group of practitioner communities, each with deep expertise in their own fields." Ken went on to say that "...for every successful event there needs to be a takeaway that changes our point of view. For me it was captured when Jason Briggs and Scott Muench of J2 Innovations took the stage and planted the seeds of change and showed us all how a connected Haystack using data modeling could alter industry dynamics as we know them." The event typically includes multiple tracks of sessions and speakers covering topics from analytics, visualization and data modeling to cyber security. An open forum filled with people who are willing to share their knowledge and experience is one of the best ways this author can think of to accelerate market change.

This has been a limited introduction to one of the most exciting and rapidly evolving movements in the industry. Again the reader is encouraged to follow developments with Project Haystack, and to make compatibility with the Open Source Data Model a requirement for every new product and system that is purchased.

# Chapter 15

# The Internet of Things (IoT)

*Glen Allmendinger,*
*Harbor Research, Inc.*

Glen Allmendinger of Harbor Research Inc. could almost be called the father of IoT. He founded Harbor Research Inc. more than 30 years ago and has worked with a who's who of companies internationally, helping them to develop strategies around emerging technology such as IoT. He and Harbor Research Inc. established IoT leadership more than a decade ago, so there is no better source for an overview of this extremely important topic.

---

*We have entered an era where people, businesses and social organizations are beginning to understand the profound impacts awareness, collaboration, and intelligence that the Internet of Things will bring. In the not too distant future, hundreds of millions, then billions, of individuals and businesses, with billions, then trillions, of smart, communicating devices, will stretch the boundaries of today's business and social systems, and create the potential to change the way we work, learn, entertain and innovate.*

*Harbor Research, Inc.*

---

## INTRODUCTION

The Internet of Things is a global technological opportunity of unprecedented proportions. The technology behind the Internet of Things will ignite an era of smart systems that radically transforms customer service, resources allocation, and productivity in general. In this new era, manufactured objects, general infrastructure, people, and businesses will be connected and communicating constantly. As the world becomes connected, societies and governments alike will be able to respond to the needs of both citizens and assets in real time.

The internet is the profound driving force on the path to a truly connected world. First, the internet gave us e-mail and the Web, then came e-commerce and the ability of consumers and businesses to quickly purchase items, as well as the ability of retailers and businesses to build close relationships based on customer preferences. But the next phase will be much more profound. It includes shared and secure internet access for device interaction—in many cases with no human intervention at all—as well as network-based services such as status monitoring, usage tracking, consumable replenishing, and automated repair. These new services are based upon the convergence of networks, embedded computing, control, content, and sensor feedback.

Device manufacturers and service companies have important roles to play, and will be the major enablers of smart products, solutions, and systems. Virtually all markets will offer new opportunities to manufacturers, as intelligence will be embedded in categories ranging from machinery and building control systems to consumer goods.

This chapter examines the current state and evolution of the Internet of Things. It also provides a perspective on emerging opportunities enabled by intelligent device networking. More importantly, it outlines how adopters are using device networking solutions, and how device manufacturers and related service companies should leverage networking technologies to gain new operational and competitive advantages.


DEFINITIONS AND CONTEXT

Since the beginning of computing there have essentially been three waves of technology and architecture: mainframe computing, personal computing, and network computing. This next generation of IoT technology will add significant new capabilities to computing and network systems. These new capabilities will revolve around real-time situational awareness and automated analysis. As a result, technology moves beyond just proposing task solutions—such as executing a work order or a sales order—to sensing what is happening in the world around it, analyzing that new information for risks and possibilities, presenting alternatives, and taking actions.

**Definition**: the Internet of Things and smart systems are a new generation of systems architecture (hardware, software, network technologies and managed services) that provides real-time awareness based on inputs from machines, sensors, people, video streams, maps, newsfeeds and more that integrate people, processes, and knowledge to enable collective awareness and better decision making.

The three previous waves of technology each have had significant impacts on productivity and efficiencies: 1) Mainframes standardized transactions; 2) personal computing placed processing power into the hands of professionals; and, 3) networked systems enabled business process automation. What is important about this next wave of smart systems is the combined impact of the cycles of innovation. While there is standalone value in each of the innovations in software systems, server infrastructure, network infrastructure, and client devices, it is the combination of all these innovations that will allow computing technologies to inform smarter systems.

As networks have invaded the "physical" world, traditionally unique components and interfaces between and among electronic, as well as mechanical elements, are becoming more and more standardized. The implications of these trends are enormous. No product development organization, nor its suppliers of components and sub-systems, will be able to ignore these forces—product and service design will increasingly be influenced by common components and sub-systems. Vertically defined, stand-alone products and application markets will increasingly become a part of a larger "horizontal" set of standards for hardware, software and communications.

As it becomes easier and easier to design and develop smart systems, competitive differentiation will shift away from unique, vertically focused product features, to differentiation based on how the product is actually used, and how the product fosters interactions between and among users in a networked context. This opens nearly infinite opportunities for forward thinking product and service organizations. Businesses can begin to explore many new possibilities for system solutions unthinkable just a few years ago. The possibilities that IoT offers for energy and analytics, as discussed throughout this book, and combined with the developments in electricity and energy markets are particularly exciting.

TRENDS DRIVING THE INTERNET OF THINGS

Harbor Research Inc. expects the rate of investment in the Internet of Things and smart systems to be measurably higher than in maturing

IT systems and network infrastructure and that investment will occur in five key areas:

- Smart system platform technology to integrate systems and applications

- Purpose-built device & hardware [sensors, etc.] innovation to support new user experiences

- New generation of device data management and analytics tools

- Business process integration

- Value-added application services

Beyond mobile phones, laptops and other traditional IT devices connected to networks, the scope of what is considered an "intelligent" device is expanding, as more and different products get connected to the network: Health monitoring devices on the body, telematics in vehicles for safety, automated service scheduling, and driver convenience as well as smart grid equipment and meters that can monitor real-time electricity usage—all delivered via the internet. We are entering an age of enhanced asset awareness, where real-world assets (e.g., HVAC systems, shipping containers, shop-floor machinery, pipelines, etc.) can monitor events, execute rules, and automate business processes. Intelligent device communications is converging with IT and network infrastructure to enable vast improvements in efficiency and productivity.

The technology and market forces that are driving business and government interest in smart machine to machine (M2M) and Internet of Things solutions are many, including:

- The cost of devices will continue to plummet with advances in silicon, packaging and integration technology making them network fluent and affordable to virtually everyone;

- A ubiquitous broadband infrastructure (significantly ahead of most predictions) has been developed, which is enabling virtually everyone in the developed world to be connected both to other users, to new types of sensors and instruments and to massive processing and storing infrastructure;

- Carriers are dramatically reducing monthly fees for cellular device links—most network service providers worldwide have deployed

IP networks on a broad scale and the wider availability of 3G wireless networks is fostering market adoption;

- The storage capacity and processing power of computers will continue to grow exponentially and enable a whole new generation of infrastructure to perform an enormous variety of data management tasks—both carriers and IT infrastructure providers are working to develop software and data management platforms to support new managed services;

- The miniaturization of components and the integration of a broad range of sensing capabilities into intelligent devices will continue to provide a variety of features that support the integration of digital information and sensory inputs from the physical world, thus broadening the range of possible applications available;

- The continuing development of IT systems, the internet and next generation telecommunication network architecture will enable movement and analysis of very large amounts of information (much beyond traditional communication and messaging) which will extend the value of managed services;

- Data communication standards being developed by a wide range of bodies including professional societies/standards bodies, government sponsored organizations like the smart grid interoperability panel (SGIP) and others are making data from new and legacy (older generation) systems available to a web service and to application platforms;

- Finally, technology advances have improved and simplified system and user interfaces and interactions which, in turn, are driving adoption of new business models.

Tied to these market and technology forces are a number of regulatory and economic stimulus efforts such as smart grid and healthcare initiatives that are also fostering increased market adoption. End customers participation in the design, development and management of smart systems will rise during this next period of market development:

- End users will emerge as a force and place greater emphasis on vertical solutions that readily integrate with enterprise systems, innovative solution design and more effective service and support.

- Demand for adaptability, agility and features will grow. Innovation in product and systems design will be heavily rewarded. Customers will creatively apply and integrate technology in their work and personal lives to unimaginable levels.

- Customers will require a tailored experience from service providers. Not only do customers expect suppliers to anticipate their specific needs, they will want suppliers to project an experience for them—smart systems are driving home the importance of user experience.

- Overall, the decline in connectivity cost, coupled with declining device costs, is driving growth in virtually all device communications. This trend, plus substantial regulatory investment (e.g. smart grid) and emergence of new features from network operators and IT infrastructure suppliers, is driving the connectivity trend.

- The Internet of Things needs a universal alternative to the many existing techniques for connecting ordinary devices to the internet. The other techniques all have something to recommend them; each is optimized for a special purpose. But in return for their optimality, they sacrifice compatibility. Since most device connectivity rarely requires maximum optimality, compatibility is a much more important objective. IP is the only answer.

- The technical development in low-cost sensors and actuators combined with low-power communication technologies such as IEEE 802.15.4, low-power Wi-Fi, and power line communication has accelerated in the last few years. However, the biggest problem with the evolution of wireless sensor networks has been the large number of proprietary users.

Many IT equipment suppliers and wireless carriers have made recent pronouncements concerning the scale of the Internet of Things. Harbor Research Inc. analysis points to the many significant challenges in realizing such growth:

- Challenges in adopting new business models and making the business case to support investments.

- Complex services and solution delivery eco-systems that require businesses to relate in new and different ways.

- Anticipation of new product, service and systems innovation modes that are not widely adopted today.

- Fragmented M2M and smart systems vendor landscape that is not yet well aligned with the larger IT infrastructure and carrier players.

- Requirement for vertically focused solutions from a supply-side world that historically has been far more horizontally driven.

## IoT Enables a New Generation of "Horizontal" Software to Enable Vertical Apps

Once a device becomes networked and is monitored for the primary purposes of device status, usage tracking, and consumables replenishment, it will also serve the larger business purpose of being a key driver for the vertical customization of services in general. For example, "asset management" is an important service that incorporates a number of different variables and systems (diagnostics for equipment health monitoring, location services for maintenance and spare parts planning, etc.). Application service providers need to organize the devices and system capabilities they offer configured for the environment in which they operate—factory, office, hospital, and elsewhere. A product inventory program will have a much different configuration in a factory than it will in an office building. More than ever before, the drivers, needs, and environmental conditions will determine the way technology is implemented. Ultimately, all devices and services, like asset management, will be highly configurable to match the needs of a particular arena, or even of a particular end-user. Harbor Research Inc. historically referred to these foundation/platform functions as "systems applications." This means a set of state-based application functions that are horizontal in nature and often characterized in a general sense as "middleware" that are not seen by the end user (we do not like the term middleware, as it risks generalizing capabilities that are unique to the pervasive internet and offer unique opportunities for suppliers).

In its most basic and practical form, Harbor Research Inc.'s concept of "systems applications" is based on "managed services integrated with embedded computing." But that's not as simple as it sounds. Capturing the real value of internet-connected devices goes much further than providing connectivity/interoperability, data basing, and some XML-based transport scheme. For example, real pervasive managed services will allow networked, embedded devices to execute remote applications as if those applications were part of the internal operating system. This type of enablement can bring extraordinary value to the growing population of network-embedded devices.

System applications are fairly generalized and are created by applying generic connectivity functions to a particular arena. The breakdown of system applications is as follows:

- **Status, Monitoring & Diagnostics**: Status applications capture and report on the operation, performance, and usage of a device, or the environment that the device is monitoring. Diagnostics applications allow for remote monitoring, troubleshooting, repair, and maintenance of networked devices.

- **Upgrades & Configuration Management**: Upgrade applications improve or augment the performance or features of a device. They can prevent problems with version control, technology obsolescence, and device failure. This kind of program makes site visits to upgrade products unnecessary and eliminates the need to keep track of what has been upgraded and when, thus saving time and money.

- **Control & Automation**: Control and automation applications coordinate devices into a sequenced pattern of behavior. These applications also allow for special-case discrete actions of a device under certain circumstances.

- **Location & Tracking**: Profiling and behavior-tracking applications are used to monitor variations in geography, culture, performance, usage, and sales of a device. Such applications can also be used to create a more customized or predictive response to end-users of a device.

- **Data Management & Analytics**: Business intelligence and specialized analytical software such as data mining and predictive analytics, enterprise energy management, ESCO 2.0, video image analysis, pattern recognition, and artificial intelligence algorithms.

### Analytics Drives Real-time Value

"Real-time awareness" is driving renewed interest and deployment of analytic tools. Analyzing and storing the massive amounts of data that will be received is only possible with extensible and adaptable systems. Rules engines and work flow are the existing technologies for deciding which alternative courses to pursue, either automatically through the application of a rule that says, "if this happens, do this," or through human review based on work flow engines that route the anomaly and alternative courses of action to the right person to make a decision. The basic function of rules engines and work flow will stay constant—seismic leaps will be necessary in the data flow and analytical inputs in a world of vastly expanded real-time awareness.

The conversion of process applications to service-oriented architectures will allow these process apps to be adapted to business scenarios, with specific components pushed down to intelligent devices where they can execute a specified action. For example, alerting a citizen on her Smartphone to the updated arrival time of a bus that was stuck in traffic, notifying a doctor on a tablet device about the drug allergies of a patient he is about to see, or directing the thermostat in an individual home to raise the temperature by turning down the air conditioner set-point by three degrees, perhaps for demand response.

### Value-added Application [Managed] Services

An important characteristic of the smart systems opportunity will be vertically focused solutions—the bulk of which will increasingly be delivered as managed or value-added services. In conceptualizing how platforms would support managed services, system-level applications would be called upon and integrated in differing configurations to provide vertical value-added applications. For example, a combination of monitoring, diagnostic, control and tracking functions could be configured to provide basic functionality required to enable an energy management application.

Value added application services are solutions that integrate people, business processes and assets and are delivered as managed services. Examples of value-added applications services include the following:

- Asset management and optimization

- Supply chain integration & business process management

- Customer support

- Energy management

- Security management

Companies will create solutions that combine elements of industry-specialized hardware devices, vertical industry software, and industry-focused wireless/wired networks with industry-oriented analytics to optimize business processes and performance both operationally and financially. Smart systems technologies and applications will help organizations address the key challenge of optimizing the value of their balance sheets, allowing them to move beyond financial assets and liabilities to their physical assets and liabilities (like electric grids or hospitals) and then to their intangible assets and liabilities (like a skilled workforce or brand). Assets and liabilities tend to be very industry-specific, even more so than processes that may be common across industries. And the task of optimizing the value of these assets and liabilities is vertically focused because optimization requirements and goals vary dramatically from industry to industry.

**Potential for New Innovations to Impact Growth**

There is now substantially greater recognition of the technological capabilities and the potential benefits of connecting devices to the internet than there was even a few years ago. This represents a whole new generation of technology innovation and, if history repeats because certain conditions repeat, Harbor Research Inc. expects a significant wave of growth in smart systems across the entire economic landscape. As Moore's law takes over and the price of embedding intelligence and connectivity into devices continues to fall, networked devices will push farther and farther into the mainstream. This process is somewhat self-reinforcing as low prices are driven by high quantities, and vice versa, making these devices increasingly prevalent in human activity and businesses. Applications will spread through all areas of life. As technology, business and supplier solution challenges are met, immense growth will begin to really accelerate.

Harbor Research Inc. believes there are indications that this era is fostering new types of innovation based on new technologies that will impact the growth of IoT and smart systems. There are several phenomenon that can potentially drive significant new modes of innovation that will positively impact adoption and growth. To understand the future of networked information, it helps to remember that the Internet of Things will largely be driven by collaboration and collective intelligence. Some important forces evident in the marketplace include:

- **Collaboration and Crowd Sourcing Are Real**: It is becoming increasingly clear that "crowd sourcing," meaning that a large group of people can create a collective work whose value far exceeds that provided by any of the individual participants, including building applications that literally get better the more people use them; and, designing participatory systems that harness network effects not only to acquire users, but also to learn from them and build on their contributions. From Google and Amazon to Wikipedia, eBay, and craigslist, Harbor Research Inc. is seeing that value is facilitated by technology, but was co-created by communities of connected users. This trend towards collaborative systems continues with newer platforms like YouTube, Facebook, and Twitter, and, Harbor Research Inc. believes, will begin to spill over into new and innovative smart systems.

- **Smartphones Are Driving Innovations in User Experience and Sensory Systems**: Many people fail to understand the significant sensor innovations that are being designed into smartphones. Today's smartphones contain microphones, cameras, motion sensors, proximity sensors, and location sensors. New internet-connected GPS applications also have built-in feedback loops, reporting your speed and using it to estimate arrival time based on traffic patterns that are no longer estimated but measured and reported in real time. Sensor-based applications can be designed to get better the more people use them, collecting data that create a virtuous feedback loop that creates more usage. Sensors are dramatically improving the user experience.

- **Analytics Drive New Values**: Data analysis, visualization, and other techniques for seeing patterns in data are going to be an increasingly valuable. Sensors and monitoring programs are not acting alone, but in concert with their human partners and an increasing number of machine learning algorithms. The continuing promises of an Internet of Things will finally produce tangible value via a hodgepodge of sensor data contributing, bottom-up, to machine learning applications that gradually find more patterns and make sense of the data that are handed to them.

- **Information Signatures Drive Awareness to New Heights**: As more and more microcontrollers and sensors are embedded in everyday objects, encoded information in physical objects will also create pervasive information signatures. Seen in this way, a printed bar code, a CD or DVD disc, a house key, or even the pages of a book can have the status of an "information signature" on a network. A product on the supermarket shelf, a car on a dealer's lot, a pallet of newly produced food sitting on a loading dock— all have information signatures now. In many cases, these information signatures are linked with their real world analogs by unique identifiers: an ISBN or ASIN, or a part number. Take the smart grid as an example—neutral web services back ends for energy-related sensor data will combine smart meter and power device data from homes and businesses to discover unique energy signatures—it is possible to determine not only the wattage being drawn by the device, but the make and model of each major appliance or any electrical device plugged into any wall socket. Signatures will combine with data fusion technology to drive rapid advances in systems awareness.

- **The Internet is Becoming A Real-time Medium**: Applications driven by machines and sensors demand that data are collected, presented, and acted upon in real time. The web is no longer a collection of static pages of HTML that describe something in the world; it is becoming a real-time communication and information forum for interactions. The internet is getting faster than you might think and real time is not limited to social media. Much as Google has realized that a link is a vote, retailers like Wal-Mart realized that a customer purchasing an item is a vote and the cash register is a sensor counting that vote. Real-time feedback loops drive inventory and availability. Sensor-driven purchasing and real-time analytics is having a huge impact on the business world and customer response.

- **B2C Drives B2B and Visa Versa**: The Internet of Things and smart systems phenomenon is not limited to consumers with smartphones. Cisco's Connected Cities initiative and its "planetary skin" project with NASA, as well as IBM's Smarter Planet program demonstrate how the B2B world is being transformed by the sensors on the internet. Factories, refineries, steel mills, and supply chains are being instrumented with sensors and machine analytics that we see in mobile consumer applications. The Internet of Things and People will depend on managing, understanding, and responding to massive amounts of user and machine-generated data in real time. With more users and sensors feeding more applications and platforms, innovators and developers are able to tackle serious real-world problems. As a result, the smart systems opportunity is no longer growing arithmetically; Harbor Research Inc. believes it's growing exponentially.

## BUILDINGS OPPORTUNITIES

The buildings arena includes HVAC, security and access, lighting, fire, people-moving systems and safety systems, as well as the energy they consume, that reside in facilities in commercial, institutional, industrial and residential segments. This section focuses on the applications and untapped potential enabled by Internet of Things technologies in homes, businesses, and factories. Companies can use the captive information in these products to improve customer service and marketing efforts, provide remote diagnostics and customized services, optimize their own product development, and increase arena.

### Evolution of Smart Building Applications

It is clear that customers seeking building automation and control systems are beginning to demand wider integration of multiple systems within a building. Therefore customers are also demanding IP-based platform architectures for these systems, as a prerequisite for integration. IT architecture in general, and IP-based device integration specifically, combines to create a common schema towards which all building systems are converging, whether they are traditionally non-IT-based systems like HVAC, security and lighting, or native IT systems like IP telephony or a computer-to-computer data network. Key trends in the buildings arena include:

* Energy management continues to drive much of the interest in the smart buildings arena. Enterprises facing increasing pressure to cut costs and reduce environmental impact are turning to smart buildings as a means of reducing energy requirements and increasing the bottom line. In the residential market, homeowners are looking to cut costs and be more "green." Demand response systems for residential users have traditionally lagged behind commercial implementers but this trend may change with increasing investment and government support for smart grid infrastructure.

- In recent years, sales of IP-based building control systems out-numbered sales of proprietary non-IP building control systems. The move toward open IP-based building automation and away from proprietary systems continues to be an important trend as open systems permit increased interoperability and connectedness between devices made by different vendors.

- Convergence between internet-enabled IT systems and all other building automation systems (BAS), including telephone, data, video, safety, fire alarm, digital signage, energy, environmental systems, and building maintenance monitoring is clearly picking up momentum. There is pressure to be able to use open protocols to mix and match (plug and play) equipment from various vendors and generations.

- Wireless building control systems are reducing the disruptiveness that retrofits and smart building infrastructure installation entails. For continuously occupied buildings (like hospitals, apartment complexes and hotels), wireless sensor networks can be implemented without the difficulty that installing wired systems entails. Removing this barrier will continue to have a positive impact on the adoption of smart solutions in the buildings arena. Additionally, wireless technologies allow facilities managers to seamlessly add new devices and reconfigure existing systems as needs change, which is driving the growing use of many new types of sensors.

- Building automation systems are being used to monitor critical assets within facilities like server rooms, critical hospital infrastructure, and security systems. Downtime among these critical assets is unacceptable, and their health is affected by environmental factors such as temperature, ventilation and humidity. Smart monitoring solutions can mitigate these threats and ensure uptime of critical assets around the clock.

- Home automation, long predicted by futurists but slow to occur in reality, may finally begin to pick up speed as new technologies continue to penetrate homes. Smartphones and tablet computers are poised to fill a void as an inexpensive and user-friendly "universal remote." Homeowners will be able to monitor and collect data about energy usage, which lights are on, which windows are open, and whether the security system is functioning properly from anywhere in the world.

**Today's Marketplace**

Harbor Research Inc. estimates that as many as 40% of the new buildings could be IT-BAS convergent or intelligent over the next 8-10 years and that the retrofit rate for existing buildings may exceed 10-15%. Meanwhile, Harbor Research Inc. expects sensor demand to grow briskly over the coming decade.

The large incumbent energy system control companies such as Honeywell, Johnson Controls, Siemens and Schneider Electric and HVAC concerns such as Trane and Carrier Corporation have all developed in the past 2-3 years open protocol solutions to complement their legacy, mostly proprietary, product lines and they are now starting to offer wireless solutions as well as demand response. Some of these suppliers are expected to expand into security system applications and then propose full building intelligence systems, more comprehensive building O&M contracts and get involved in analytics technology as well. In parallel, many new players are addressing the emergent "energy intelligence," or energy and analytics, arena as it is central to the concerns of building operators. Increased building intelligence will not be restricted to new buildings (where it may be easier to sell) but it will also become increasingly applicable in existing buildings. Retrofitting an existing building is already doable with current emerging technology. Furthermore, many retrofit improvements can occur without any local utility involvement.

Increased building intelligence will mean more precise and open energy and power management in commercial premises as a result of being able to relay on demand real-time precise point-of-use data to web portals that will be openly accessible by building guests, tenants and owners alike (subject to security rules). On-demand remote metering and monitoring of any load will be possible. So, for the first time, commercial electricity management will be an open playing field where building managers, owners and tenants will all be able to truly "discover" and manage their energy needs as they want. Owners, managers, and occupants will have the ability to access precise real-time information on demand as well as the ability to retrieve information remotely via mobile phones and computers.

Open commercial energy management will not only benefit the

building engineer or manager, it will also reach the building owners' CFO, CIO, and operations executives who can now oversee large commercial real estate portfolios on a regional or national basis.

Tenants in intelligent commercial buildings will also be able to monitor and influence their demand profiles and routines to truly control their energy use, level of comfort and energy bills, all at once. This will be much better than receiving a monthly statement based on the amount of square feet they happen to rent (especially if some arcane adjustments have been made by cost accountants that do not necessarily have any idea of how the building operates and what various types of equipment different tenants use or how these tenants use their space).

This will completely change the game; more stakeholders will have a true seat at the decision table; these new constituencies will have to be recognized and catered to; new business propositions will emerge; and the marketing of energy will change. Tenants will be able to relay their site information to their headquarters which will be able to quickly aggregate for regional or national enterprises including, of course, the opportunity to shop for electricity supply. For example, building intelligence will help solve the tenant conundrum. Right now, tenants generally pay for their energy as a more or less transparent part of their rental lease payment. With building intelligence, tenants can be sub-metered (and meter any load they want) and take local, precise and cost-effective actions that affect not only the entire premises but also select parts of the rented space. This will be the end of the blind energy lease charge, the "one-size-fits-all" building management approach and the end to the tons of excuses that building owners can throw at tenants to offer them half-baked approaches.

Increased building intelligence (and the underlying IT/BAS convergence that it implies) will also enable or allow increased or better use of many new energy saving and demand-responsive technologies such as:

- Daylighting and fully programmable (luminescence-based) lighting (based on the DALI standard) will be prevalent in new buildings past 2010.

- HVAC efficiency was projected to increase 5% by 2010 over previous building stock, and around half of new buildings were expected to adopt new demand-control ventilation techniques and 20% to adopt passive cooling properly integrated in the building envelopes.

- Innovative demand-controlled ventilation designs.

- Decentralized DC (Direct Current) UPS networks in data centers.

- Packaged plug-and-play DG.

The payoff will be huge: tenants will be able to buy their energy from their own supplier in states that allow this. They will also have the freedom of participating in an individual customized or building-wide basis in local demand response schemes based on their specific supply-demand profile. Buildings will become a major source of demand response. Data show that, even though it accounts for 35% of the total US load, the commercial sector accounts for 45% on average of summer electric peak coincident demand, with a sector peak in excess of 330 GW. The ability to precisely trigger on-demand DR could help reduce the sector's peak by at least 10-15%, based on various DR experiments conducted to date. It will become possible to enroll and aggregate thousands of buildings in DR programs. The impact could be considered if we believe some estimates that show a 2.5% reduction in peak demand could reduce the costs of serving that peak by 25%.

Tenants will be able to decide whether to invest in dedicated UPS, storage or distributed generation, or rent these capabilities from the building central manager or potentially from adjacent or nearby facilities. If they do, the UPS, storage and DG systems will also be fully web-enabled and managed. There is growing use of decentralized DC-based UPS systems; smart mid-size grid-connected "active" two-way storage technologies; and the ability to deploy scalable plug and play DG (including better DG packaging and better interconnection modules). In parallel, the deployment of automation and integration technologies will usher the entry of next-generation maintenance management programs, allowing equipment condition monitoring, early fault detection, and predictive and self-healing maintenance approaches. Operating personnel will be effectively dispatched through wireless communications through their PDAs or cell phones.

Finally, one can also progressively expect a new way to design and develop new buildings, if only because the buildings of the future are likely to fulfill new functions and operate differently, especially in the office, hotel, health and education sectors. We now sense a reasonably strong momentum toward green and smart building and a better way to spec energy systems for that purpose and truly consider more long-term costing approaches. Still, many observers keep saying that the commercial real estate sector will continue to operate under an ancient and dysfunctional system where the decision maker ends up being a general contractor which is neither the end-user nor the future facility manager. As a result, the sector is served by a fragmented vendor population: architects, specialty designers (e.g., for UPS, DG), building control vendors, HVAC vendors, UPS vendors, and DG vendors.

How long the industry remains fragmented will depend on how ubiquitous new smart commercial building management systems get implemented, how quickly energy retail deregulation happens, and the speed at which DR programs get implemented.

**Adoption Progress**

The fundamental difficulty in achieving systems intelligence in buildings is communication, and this problem can be broken down into two types of instances. The first issue is compatibility. For example, because the company that makes light switches doesn't also manufacture water boilers, there is no reason these two devices would have been built to speak the same language. The second issue is connectivity—the fire alarm device from one manufacture is not connected to anything else. Today more than ever, there are a number of solutions that address these problems from all different angles. In fact, many experts would say that the biggest roadblock to wider adoption of intelligent systems for buildings no longer has to do with technology.

Regardless of building size, the overall goal of building managers is maintaining an environment that is responsive to the needs of those who work, shop, and live there. Managers are driven by concerns that can largely be organized into two segments of interest: comfort/convenience and safety/security. Of course, these concerns fit into a broader economic context which can best be summarized as, "Please do not add additional costs and, if possible, please reduce my costs as you add new values."

The benefits of interoperability are obvious. For example, linking life safety systems to other building systems can improve response in an emergency. The fire system could alert the HVAC system to pressurize stairwells for occupant escape; elevators could be alerted to drop off occupants at the safest, lowest floor; standby power could start up to activate emergency exit lighting. The key to attracting building managers or consumers to internet-enabled applications remains a low price point, easy installation, and simple systems management. A key challenge for technology suppliers and service providers is the fragmentation of the end markets. From a device supplier perspective, available channels are also relatively fragmented. In spite of this, interest and activity is quite high. The desire for simplicity and high uptime is largely driven by smaller staffs with less technical ability. In the case of asset management and support coordination, the objective is to fix problems remotely, or if that is impossible, to dispatch a person with the right expertise, tools, and parts.

Energy has really become the "killer app" of the buildings and home arenas. The market for energy measurement, monitoring and control devices crosses the industrial and commercial sectors and its size is difficult to assess. Each of these market sectors will benefit from the ability to control power and even reduce consumption permanently, thereby reducing carbon emissions. Demand response solutions today are doing this on a contracted basis in order to alleviate load during peak periods of the summer months, however they strive to implement more automated energy management solutions and/or those that can permanently reduce load. Remote management is an additional benefit which could further reduce operating costs in the commercial/industrial sectors and some high-end property and property management solutions. Today's solutions for building energy management are rapidly converging with IoT technologies and include:

- AMR/AMI (automated meter reading/advanced meter infrastructure),

- Demand response, and

- Utility sponsored/consumer energy efficiency products.

Unfortunately, some of these solutions are limited by a lack of cost-effective, space-efficient sensors to measure, monitor and manage energy demand and consumption. For instance, demand response solutions must rely on expensive, external control mechanisms to cycle-down large equipment and disconnect loads, so limiting their reach and their cost-effectiveness. AMR/AMI solutions are beginning to grant a much better view from utility to meter, but they still offer virtually no insight beyond the meter, offering little to no control to the occupant to monitor and manage electricity consumption. Moreover, none of these solutions addresses the huge, untapped opportunity to provide information and control of the commercial building to consumers. Clearly, energy efficiency efforts have begun, but they have a long way to go, and it is the last few yards (beyond the meter) that really matter in managing and reducing the world's energy consumption.

Combining the pervasiveness of networking with embedded energy intelligence enables for the first time new real-time capabilities for demand-side management even down to socket and appliance level. As concern grows around power quality, availability and uptime, these technologies provide a unique unobtrusive means to provide customers with detailed load profiles and usage information that will empower end users to take control of their energy use. By setting building policies, end users will be able to safely forget about their energy consumption while their buildings look after their interests—actively connecting and disconnecting unused and unnecessary loads automatically as they go about their daily lives. Areas like sub-billing/metering, easier servicing of equipment and the ability to upgrade "in-site" and set sensitivity parameters are attractive benefits. Ultimately, organizing load management without disrupting customer life styles will be key to attaining important "green" goals in society. Smart energy intelligence can enable industrial, commercial and home smart energy networks, which can be used to monitor energy quality, switch and control energy sources, reduce energy consumption, and save costs.

**Future Opportunities**

Harbor Research Inc. believes that the following market forces deserve the acute attention, as they will have great impact on the requirements and direction of the market for building automation systems:

- Energy efficiency—rising energy prices and government regulations are encouraging building owners to operate more efficiently;

- IT integration occurring throughout the consumer environment is driving the desire and assumption for it to occur everywhere;

- Rising cost pressures from increased outsourcing and contractor buying;

- Proliferation of IT-based service delivery platforms like XML and SOAP; and,

- Wireless technology is gaining momentum by addressing its two key drawbacks—reliability and cost.

The advent of "green" building design will also have a growing effect on the market. Most green building technologies are mature enough today that investments should not be considered high risk. The most efficient green buildings take a complete system perspective of requirements and capacities, incorporating real-time heating and cooling adjustments to reduce costs and improve effectiveness. Significant cost reduction opportunities for green projects are available for qualified investments. There is a growing interest among both traditional building equipment suppliers and IT vendors in entering the green building industry with new products and services.

# Chapter 16

# Energy and Analytics Tools

The topic of energy and analytics tools is really a culmination for this book and yet it is a daunting topic. After years of evaluating this technological development and surveying offers in the marketplace, it has become evident to this author that a comprehensive review of all the tools available in the marketplace is unrealistic. Therefore, this chapter will take the same approach espoused throughout the text; it will encourage the user to carefully evaluate system requirements before exploring technology or "tool" options. A thorough review of requirements should result in a framework/specification that can be used to consider solutions. Many building owners and operators have completed such an analysis only to realize that there is no offering currently available that fulfills all of their needs. In some cases, these operators have teamed with a company that they felt had a "best in class" offering, with the understanding that ongoing advancement would be necessary to retain the business and evolve the solution to meet all of that customer's needs. In other cases, some building owners have elected to create their own solution, and become a solution provider for their own properties as well as actively sell the solution to others. There are many ways to approach this issue, but every approach must start with a clearly delineated set of requirements and expectations. Of course the owner must also recognize that requirements change over time, and an assessment of the solution provider's willingness to respond with enhancements, upgrades and expanded capabilities should also be considered. This last requirement is very difficult to assess, but it is important for the customer to consider the corporate culture and ownership of the provider to make the best estimation possible of where there is alignment with the customer's business philosophy and a drive for excellence that will spur continuous offering improvement.

Once requirements are outlined, the next consideration is "best practices for analytics." This chapter will begin with an overarching review of energy and analytics Best Practices to frame the dialog for

267

a secondary set of customer requirements. These requirements should consider the scalability, resilience and long-term viability of solutions. As part of that discussion, the chapter will review pilot implementation steps that have been drafted by SkyFoundry. The final topic of this chapter is analytics architecture. In this architecture section, the makeup and infrastructure of analytics offerings will be reviewed. This is important because much of the coverage readers will find in the media focuses on the analytics tool/engine, rules, algorithms and visualization, but not on the actual infrastructure. Previous chapters have covered topics related to what analytics are capable of and how rules, algorithms and visualization can be deployed for energy management. The focus here will be on the actual technology implementation, so that the manager can truly understand the context of implementation, how invasive it will be, and what might happen if a break-up decision is made downstream. A manager who has assimilated this knowledge will be in the best possible position to evaluate energy and analytics tools and make an informed decision about the best solution to procure. This section will close with a brief discussion of training. Training will be covered under two headings: 1) training for all customer staff on the selected solution and 2) industry training to maintain a continuously improving energy and analytics program that results in the highest level of performance for buildings and campuses.


## ENERGY AND ANALYTICS BEST PRACTICES: SETTING THE BAR

Applying the science of analytics to energy management, particularly in the 21st century, transforms the work that must be done by building operators and Certified Energy Managers (CEMs). As has been discussed throughout this book, this is especially true for electricity which has seen a business transformation and redefinition of the supply and demand sides of the meter. Equally important to managing energy is also reliability (or "resiliency"), especially for electricity, and there are a host of other topics in the spotlight globally, besides resiliency, in this era of challenges including climate change. At the building and campus level, the task of managing energy use and cost remains as critical and difficult as ever. That is why it is exciting to see CEMs being presented with a new set of tools to help carry out these tasks. Energy and analytics have been covered from every perspective in the book thus far; yet there are other instances of this technology to consider, particularly the Internet of Things, which was discussed in the last chapter. In that chapter Harbor Research Inc. pointed out that CEMs who deploy IoT may be able to achieve more data and potentially higher levels of energy optimization than ever before. The Internet of Things is exciting and will be discussed here in the context of analytics tool configuration and data access. There may be unlimited applications for the Internet of Things (IoT), but analytics is one that is driving energy optimization results today. This shouldn't be surprising because building operations, with high energy cost, demand new approaches—in this case, Internet of Things-driven Analytics IoT. Before talking about the role of IoT within energy and analytics however, it is again critical to emphasize the importance of delineating requirements and expectations in any building.

So what is Analytics IoT? As this book has covered in detail, analytics is a term for discovering meaningful patterns in data. Incorporated into the definition as it has been discussed here is visualization via dashboards and other tools, but to drill down a bit deeper there are three "essential elements" that must be covered:

1.  Human expertise,
2.  High quality data, and
3.  Sophisticated analytics software.

In this section these topics will be discussed in more detail because they are essential to evaluating analytics tools. Treatment of BAS, data communication, middleware and Internet of Things (IoT) in previous chapters was provided because understanding these topics is essential to evaluating tools. The manager can apply this technology understanding to an evaluation of how these elements are packaged and addressed in any analytics tool. Tools are being offered to the industry that leverage the advanced interoperability of devices, systems, and services. Yet the tools referred to here as Analytics IoT apply the essential elements along with the energy and analytics technology/software to produce exciting results for energy management and equipment optimization in buildings and large campuses. Universities facing huge energy challenges are among the early adopters of Analytics IoT. About once a month the author receives an email from Jim Lee at Cimetrics, Inc., with stories about such early adopters. A recent story was about one Ivy League university that is achieving $315,000 in annual energy savings. This campus began with 3 buildings, two labs and an administrative building, comprising over 400,000 square feet. In this case the IoT component, includes "interoperability" with an energy management system (EMS) to connect to, and from which to collect over 5,200 physical points of sensor and actuator data. These data were collected from air handlers, terminal units, pumps and other equipment every 15 minutes, 24 hours a day, 365 days a year, for a total of approximately 500,000 data samples per day. The analytics component function performed over 1,000 software algorithms continuously analyze the data and identify opportunities to reduce energy use, improve environmental conditions and drive down operations and maintenance costs. The result was an Analytics IoT system that paid for itself in 0.23 years... yes that is 4 months! It also reduced the university's $CO_2$ emissions by 820+ tons for impressive sustainability performance too. This could not have been achieved without technology that is built on the essential elements of energy and analytics.

Certified Energy Managers, and the facilities industry as a whole, are demanding this type of technology, and a growing number of companies are offering options, or tools. The challenge for managers, and the focus here, is how to choose an energy and analytics tool that will deliver results. Analytics IoT solutions combine interoperability, data, software, analysis and an understanding of best practices in building operations. The key to that previous sentence is that it describes the "offering" as a service or "solution" rather than a "tool," and yet, so often in the energy and analytics world, providers only want to talk about the technology. As the manager considers an energy and analytics tool, it is important to evaluate the technology and the interoperability, but also to evaluate how the technology carries out the analytics and how a tool is interfaced with human expertise to touch all of the bases outlined as "essential elements."

The first essential element is human expertise. In fact the basis for this discussion is that Analytics IoT is a combination of data, technology and know-how. The data discussed in this book are specific to energy and buildings, and they can be accessed from a host of installed systems. Many might confuse Analytics IoT with "Big Data" and the term is valid but there are some distinctions. Big data is used for a wide variety of applications, like on-line retailing, and Wikipedia defines it as "an all-encompassing term for any collection of data set so large and complex that it becomes difficult to process using traditional data processing applications." Unlike those generic big data applications, the focus of Analytics IoT is to leverage big building data for a much more specific focus on energy management, buildings and systems which demand custom analytics software and technology. In this case it is more about understanding what the data tell CEMs about building operations and how that information can be used to develop strategies for better, more efficient, buildings. This aspect of Analytics IoT merges technology with know-how or "human expertise." Human expertise is the "essential element" of analytics that enables an understanding of what the data mean, and how to develop meaningful recommendations about what issues should be addressed. Consider a medical analogy. What if a patient is sick and told that a magnetic resonance imaging (MRI) scan must be done. What type of doctor is the patient most likely to want—an expert on MRI technology, or a doctor that understands both the technology and the patients' body, in order to understand how to use the information from the MRI to treat him or her successfully? The obvious answer is the same reason the ideal solutions for building analytics require experts, in many cases CEMs, who understand the technology and the building.

The question for managers is what should they look for in an energy and analytics tool? This section answers that question by providing a context for evaluating tools through the lens of these essential elements. The premise here is that the ideal tool combines human expertise with technology that effectively delivers the other two essential elements: high quality data and sophisticated analytics software. The right energy and analytics solution provider must deploy sophisticated technology, combining the elements above, along with world class analysts (human expertise) to adapt the technology to customer needs and provide continuous support. Analysts must be content experts, ideally engineers or CEMs, experienced with energy and buildings, as well as the equipment and environment that customers are striving to optimize. Analysts can then apply this understanding of unique customer applications, and deploy energy and analytics technology to turn data into knowledge about how to improve building efficiency and operations. The analyst plays an essential role to ensure data quality and deploy effective Analytics IoT technology.

The second essential element is high quality data, and managers begin by evaluating tools based upon two aspects of data quality: data sufficiency and data resiliency. Data sufficiency is the first step and the goal is to ensure that there are enough data available to conduct meaningful analysis. Having sufficient data is critical, but in most cases the installation cost to add hundreds of new data points to a building is prohibitive. So tool analysts and designers rely on data from existing sources including BAS or energy management systems (EMS), computerized maintenance management systems (CMMS) and meters, plus some external web-based sources such as utility databases for bill data. The energy and analytics tool must be able to build a sufficient and effective data architecture, which requires creating a wide range of expertise with technology, data communications, IT and computer systems. In many respects, this is more a question of how the tool deployment is designed than of the tool itself, but again energy and analytics is as much about the human expertise as the tool itself. However it is important for tools to be flexible in the data that can be accessed and, as will be discussed later, in the rules against which those data are evaluated. Tool providers must leverage business process experience with the type of building or campus in question, and conduct ongoing communication and coordination with client teams.

It may be helpful to illustrate the importance of data sufficiency, by discussing it in the context of one specific building type—universities. Focusing on universities and, in particular energy analysis, shows how important it is to have sufficient data. Energy metering for universities can be an issue for two reasons. First, many universities have traditionally been master metered by the electric and natural gas utilities, without building level meters, a fact that is changing. Second, universities often utilize central plants for hot and chilled water loops, so energy consumption at the building level is not easily compared with energy benchmarks for buildings that include consumption by prime movers such as boilers and chillers. Analysts must consider and identify all of the data that will be needed to conduct analysis and produce recommendations for improvement. In this case, energy may not be consumed by prime movers, but it is possible to meter the hot and chilled water loop and measure the energy content of that water entering a building as Btus. One slight issue is that this approach ignores the energy conversion losses, such as natural gas burned by a boiler to produce heat, which reflects efficiency or how well the boiler converts energy into heat. This makes it hard to compare a building with a boiler to a building served by a central plant. Yet comparing buildings on the campus to one another is very effective. This is another good reason why the Energy Star™ program provides benchmarking comparison data by building type, allowing universities to compare performance against other universities. To bring this discussion back to data quality, the point is that managers must evaluate tools based upon how data sufficiency is addressed. Tools must provide the maximum number mechanisms for accessing data needed to perform meaningful analysis from any source, underscoring the importance of leveraging interoperability with all building systems, web-sources etc. Equally important to data sufficiency is how the tool ensures that data flow or that resiliency is not interrupted.

Data resiliency is the second aspect of high quality data, and it is critical to ensuring that analytics solutions remain viable. A common threat for analytics applications involves networking and data communication. Almost any type of building today will have IT and network administrator groups that must be interfaced with, and which must give permission to implementation of energy and analytics. If middleware, such as gateways or third-party communications devices, must be installed as part of the solution, extra diligence is required, especially if that network administrator is asked to provide an internet protocol (IP) address. The manager, or analytics solution provider, will need to pay ongoing attention to the data being provided to be sure that communication, and therefore data flow, is not disrupted. IT departments and network administrators, some of whom are outsourced service providers themselves, implement "changes," almost continuously. These "changes" can disrupt data resiliency and may include building renovations that cause re-cabling or upgrades to networking routers and switches and changes to information technology (IT) and internet protocol (IP) addressing. When this happens, data flow is disrupted, and the value of energy and analytics tools is diminished. This is because the analytics tool is analyzing old data or analysis is suspended for a period of time.

There are two critical data resiliency take-aways: 1) it is important to forge robust ties between IT and plant departments, and 2) it is critical to conduct data assessments that are both technology and human driven, to ensure data resiliency. The topic points out a very important evaluation criteria that will be discussed below under sophisticated analytics software. That criteria is to evaluate how the tool operates when the data flow is interrupted; for example, does the tool keep conducting analysis with old data, does it estimate data, etc.? A final note under data resiliency is that energy and analytics tool providers, or the manager's organization itself, must have connectivity and interoperability expertise, to ensure continuous flow of reliable data. This is the single most important reason for the organization of this book. The technology section of the book has covered the full range of topics from BAS to data communications and middleware technology to emphasize how important this technology is, as a foundation for the essential elements of energy and analytics. Data resiliency is part of that foundation, and it requires a method, human or technology-oriented, to continuously monitor data integrity. This process must evaluate data access and integrity, and must also be able to create electronic alerts to notify appropriate personnel of the issue and corrective action needed.

Armed with high quality data, the next essential element for energy and analytics is sophisticated analytics software. The reader should note that there are myriad analytics technology start-ups that are promoting new offerings. So a first technology requirement should be that the offering is field proven and provides hundreds (even thousands) of sophisticated algorithms, which constantly analyze data, to identify energy savings, as well as comfort, code, regulatory, safety and equipment efficiency improvement opportunities. These algorithms must evaluate energy intensity of the building and equipment, while also providing predictive maintenance capability that has been tested and field proven. These algorithms access data, as discussed above, to determine current operating conditions and then the algorithms compare those data to rules that define either optimal operation or optimal energy performance for a piece of equipment or process. The rules used may be based on design specifications for a piece of equipment, benchmarks from high performance buildings, past performance in the building being evaluated, etc. Hence, the desire for proven technology. This is because the more robust the database of performance data in the analytics tools, the more sophisticated the analytics software. Ideally algorithms may be tailored to various building types so that facility specific characteristics can be evaluated against rules. For example, indoor air quality and ventilation requirements are very different in a laboratory, hospital intensive care unit and office buildings, so it should be possible to adapt the rules, which algorithms are compared to, for a variety of facility needs. Energy and analytics software must also provide the capability for customers to add new rules that identify recommended actions to improve the customer processes on campuses. This is particularly important because managers, particularly CEMs, tend to be highly skilled and capable of sophisticated system interactions. Strategies should be offered to use technology to react to data quality or resiliency issues, and to incorporate corrective actions to compensate for missing data, in real time. If data points are missing, alternative algorithms and calculated values are utilized to perform analytics. This ensures that results can be achieved under varying levels of data availability while data resiliency issues are addressed.

Another sophisticated analytics software feature is incorporating effective software planning tools to track and prioritize recommendations for correction and improvement. Correction, or corrective action, may be a repair, etc., while improvement could be an optimization to reprogram a BAS control sequence or a retrofit that could include an equipment upgrade—i.e. adding a VFD to a fan. This is an important analytics differentiator; some systems simply list a number of issues or anomalies. More sophisticated systems combine that technology with direct interaction by an analyst who evaluates all building systems and then meets with managers to review analysis of the energy/cost savings from each recommendation and prioritize which ones to address first. The best case is for the software to track all of the corrective action/improvement recommendations, report on them, identify prioritization for addressing those items, and to reprioritize recommendations based upon new information. Even more valuable is tracking how long it has been since the recommendation was originally made and to provide a means to review and evaluate all of the recommendations. These reports can also summarize the value of savings or avoided cost associated with each recommendation, and potentially information such as how much "avoided cost" has been incurred as a result of not addressing the issue immediately. This could be called a "do nothing" evaluation. Considering the cost of doing nothing, and its impact on the potential savings, makes it easy to decide which recommendations to address first.

Applying the energy and analytics essential elements to selecting and deploying this technology cannot be understated. Analysts provide human expertise for engineering technical support and ongoing communication with management to ensure proper and timely implementation of recommendations. Customers, or managers, benefit because they become an analytics team with the analyst. The analytics team is continuously working together to fine-tune buildings and to incorporate what the analyst/manager team has learned about how specific types of facilities operate. Addressing the essential elements also improves the value of benefits that are quantified, including but limited to energy & O&M savings calculations. Only proven sophisticated analytics software coupled with human expertise can ensure a design that maintains high quality data, and can be customized to the unique characteristics of any building type. There is one additional topic that must be addressed regarding the evaluation of analytics tools, and that is analytics architecture.

### Analytics Implementation Options

To reset the context for this discussion of analytics tools, it may be helpful to frame this dialog based upon the technical content provided throughout the second half of this book. Yet the scope of the content can be daunting, so another approach might be to view analytics tools through the lens of requirements to execute an analytics pilot project. A valuable resource for accomplishing this is provided by John Petze, partner at SkyFoundry and co-founder of Project Haystack. Mr. Petze provides a resource to frame the dialog using the *SkySpark Pilot Process Overview* published on August 11, 2013. The reader may also find a more detailed overview of getting started with deployment of SkySpark software in the deployment chapter of the SkySpark documentation, http://www.skyfoundry.com/doc/docSkySpark/Deployment. These documents point out that some of the "most successful projects start with a pilot implementation." In the context of analytics tool evaluation, that is a useful point because the complexity of an implementation can be overwhelming, as can the cost. Therefore it may be helpful to think about this in the context of a pilot project. It may also be helpful for the owner to take that approach in evaluating tool options.

This pilot project document goes on to state that "pilot projects may focus on a little bit of data from a large number of sites or a more thorough application of analytics to all major equipment in a few facilities. In either case a pilot involves using real data and an implementation effort is required." The key facets of an implementation effort include:

1.  **Data Access.** Identify the source(s) of the data to be analyzed and answer the following questions:
    a.  Where is the data located?
    b.  How will you connect to the data?
    Typically the data include historical data from a BAS and/or metering system. It is also important to have basic facility data such as address, size (square foot or meters), equipment listing. The goal is to identify the available data, where it is, how you will connect and what format it is in.

2.  **Data Modeling**—also known as tagging or semantic modeling. Data must have meaning in order to be interpreted by analytics. The data on-boarding process involves tagging the source data to define meaning. The process typically starts with a detailed list of "points" of data that the manager has identified for recording. A point is anything we want to collect data for, and it can be physical or virtual. That could include sensors, control points, schedule status. The "points" list would typically include point names, descriptions, and units.

    The list would need to be annotated to describe relationships—such as "this zone temp is fed by this AHU," and "this supply air temperature sensor is measuring supply air from this AHU." In the data modeling step, the manager is defining basic information about the units, meaning, and relationships of the points that the energy and analytics tool will analyze.

3.  **Analytic Rules/Algorithms**. With an understanding of the data available that has been identified for capture, a whiteboard session is usually done to discuss and define (in English—not programming code) the types of patterns/issues to look for, such as:
    — Economizers open when they shouldn't be;
    — AHUs, lighting and other equipment that operate outside of schedules;
    — Deviations from kW benchmarks or limits we want to track;
    — Energy performance correlated with building characteristics;
    — Short cycling;
    — Simultaneous heating and cooling;

    A good pilot project typically defines 6-8 rules to start with. Start with a manageable effort to get results quickly.

    To design the energy and analytics pilot project for a tool that the manager wants to evaluate, it will typically require:

    **Human Expertise**—there are several tasks:
1.  Data: this might be the manager themselves or a CEM or project manager to lead the effort to get the above information.
2.  Systems: a systems expert to design the connectivity required to get the data. For example it will be necessary to understand how

to connect to the specific system that holds the historical data. When direct connection to a BAS is involved, this often requires some involvement from the systems integrator that implemented the system—for example a systems engineer familiar with the BAS system.

**High Quality Data Access**: In a case where the data are already in some type of database (i.e. SQL) then it will be necessary to have expertise/understanding of the database schema. This will require an understanding of the systems that the data come from and the ability to answer questions like:
— What do the point names mean?
— Which points are related to which equipment systems?
— What is the expected operation of the equipment systems?

**Engineer/CEM knowledgeable with the analytics tool** to establish the data connection (to a SQL DB for example), write the rules and determine algorithms that will be carried out.

In some cases the energy and analytics tool provider may not provide implementation services directly. This means that the manager must perform a secondary analysis of the supply channel partners, to secure the services of a company, typically a system integrator, that will provide implementations and support services for that tool.

This chapter has covered two critical topics that are very useful in selection of an energy and analytics tool: 1) best practices, and 2) the requirements for a pilot project. Once the manager has developed a set of requirements and plan, based upon this information, the next critical topic is the technology, or analytics tool itself, which will be discussed based upon the analytics architecture.

## Analytics Architecture

The topic of architecture and systems has been a recurring theme throughout this book. In the evaluation of energy and analytics tools, architecture is also a critical topic. This section will cover a number of topics to provide the reader with a framework to use in developing an approach to understanding the tools themselves, and to comparing tools against one another. As part of this process the manager must also evaluate the technical requirements that may be impacted by selecting one tool over another. This content will be addressed by covering the following topics:

— Cloud and web services solutions anywhere, anytime, any device;

— BAS, middleware and device standards and solutions;

— Dashboard and API;

— Software as a service; and

— Cyber security.

## CLOUD COMPUTING

Evolution of energy and analytics technology has been enabled by numerous advancements, but perhaps none more significant than cloud computing. The first article that this author wrote on the topic of analytics, then referred to as building simulation and analysis, was 30 years before the publishing of this book. At that time, the state of the art was to temporarily deploy a mini-computer in the building, outfitted with what was then an astronomical amount of data storage capability. That mini-computer was necessary to support the huge hard drive for data storage, but also to support the computing power necessary to run the algorithms. At that time the mini-computer was also necessary to physically, but temporarily, support a network containing hundreds, or even thousands, of sensors because there was no BAS standard like BACnet™. The cost of such simulation and analysis equipment and services made it inconceivable to consider deploying that kind of computing power in any building on a permanent basis.

Fast forward to today when it is possible to deploy web-based, cloud computing solutions, and interface with them at anytime from anywhere. Leveraging cloud-hosted energy and analytics solutions means that servers can be deployed online with the computing power for the analysis, and that the data storage can be scaled online and shared across numerous buildings. The internet connects those buildings to energy and analytics in the cloud, and can utilize a robust system with almost unlimited data storage capacity to scale with client needs. Cloud-based approaches can react faster to data quality or resiliency issues, and to incorporate corrective actions to compensate for missing data in real time.

## SOFTWARE AS A SERVICE (SaaS) OR
## DATA AS A SERVICE (DaaS)

The topic of cloud computing leads naturally to consideration of how services are priced. This book is about understanding the technology, so this topic will not be treated in any detail. It is simply important to note that, in the past, software was sold in a "shrink wrap" packaging format. The customer purchased the software, signed a license, and it belonged to them and could be used without any other costs, until an upgrade was offered at a later date. Yet products that require cloud computing, and full range of features discussed in this book, cannot be shrink wrapped, thus a new method for product offering was created call Software as a Service (SaaS). Anno Scholten, President of Connexx, has also introduced a variation on this concept in such offerings Data as a Service (DaaS), which reinforces that these approaches to technology sales are in a continual state of flux. Regarding SaaS, the intent here is not to outline the pricing models in any detail, but rather to emphasize that managers should understand how the tool is packaged and priced. This information along with an understanding of how a suite of offerings may be combined to create the best solutions for a facility's needs, are extremely help in identifying the best tool for a specific customer.

## BAS, MIDDLEWARE AND HVAC-CENTRIC SOLUTIONS

Managers who are selecting tools will note that energy and analytics solutions are being offered by a host of companies. It is important for the evaluation to begin by separating the analytics tool from the rest of that company's products and services for independent analysis. Bundling is a popular approach used by companies that combine many solutions and offer them as a solution. The first thing that may become evident by conducting independent evaluations of applications, is that some companies feel compelled to have an analytics offering, and have decided to OEM (original equipment manufacturer provided) or "white label" another company's tool. This is not a bad thing, but it will help the manager decide whether the evaluation should consider a "white label tool," or eliminate that company from consideration. In the alternative, the manager may decide to consider the tool because they believe that the company's bundled offering, including the white label analytics, brings added value that makes it desirable to acquire the tool from them. The other point that may be of interest is that some HVAC manufacturers may be implementing analytics tools specific to a piece of equipment. In this case, again the manager would want to evaluate whether to implement that company's equipment-specific analytics tool, or to integrate that functionality into a more comprehensive tool. As outlined in previous chapters, highlighting BAS and middleware, the ideal energy and analytics solution will leverage standards for networking, data modeling, etc.

## DASHBOARD AND API

One critical component to user friendly energy and analytics tools is "visualization." As outlined in the earlier chapter on visualization and dashboards, this technology presents an essential window into the analytics, as well as the primary interface through which to review recommendations and reports. As noted above, the dashboard can also provide the primary vehicle for prioritizing when to address analytics recommendations. Equally important to this discussion is the topic of an application programming interface (API). APIs were discussed in the chapter on middleware and are an approach to creating "glue" to enable interface between existing systems and a web services offering. It is important to understand and develop "requirements" for the analytics tool, so that the manager can determine it is necessary to enable a secure interface for enterprise applications allowing access to more of the data found in a BAS, etc. Which is a logical segue to the last topic under this heading cyber security.

### Cyber Security

The 21st century marketplace utilizes the internet and web services to offer a seemingly infinite number of offers to the market. Cyber security is a topic that is inseparable from any such service. This section will not outline how a strategy might be developed for cyber security with energy and analytics tools. Rather it will make a simple point: managers should evaluate all tools in the context of existing corporate or organizational policy relating to cyber security. Such an analysis will ensure that tools meet all usual and customary customer expectations for cyber security, while providing optimum service and performance.

# Chapter 17

# Financing Energy Management For Buildings and Campuses

*Co-Authored with Leighton Wolffe, Northbridge Energy Partners*

This chapter is expanded from an article published in *Engineered Systems Magazine* in June of 2012 by Leighton Wolffe and myself. Mr. Wolffe is a thought leader in the energy technology space with three decades of experience in direct digital control, web services and demand response. He is a partner in Northbridge Energy Partners (www.northbridgeep.com). Our goal with this article was to provide insight on the many energy related services and programs to help finance capital improvement projects and infrastructure upgrades, to achieve high performance buildings. This topic is relevant to energy and analytics because the methods of delivery for these products are diverse, and coupling analytics with an energy retrofit can be very effective. It is also relevant because the outcome of energy and analytics technology will be recommendations for operational changes to the building, but inevitably these tools will drive capital programs as well. This book has not been about the sales and marketing side of energy and analytics or, in that context, a study of pricing models and market channels for the sale and service of this technology. However, this limited treatment of the financing topic is relevant as a general overview of some options, other than self-funding, that exist for building owners to deal with the overarching capital needs they will have as they pursue higher performance, energy efficient buildings.

For facility professionals every mechanism to make capital improvement projects a reality is worthy of attention. However, this has become a rather complicated formula to master, with the array of vendors, energy suppliers, service providers, demand response companies, dashboards and analytics programs, and building automation system (BAS) manufacturers all claiming to offer critical components and comprehensive service for improvements to operations and energy efficiency. While all of these offers can bring value, it is important to begin a building self-evaluation by asking questions such as:

- Which project first?

- ECMs—how to prioritize?

- How do I finance?

- What if I choose a dead-end technology?

- How does this all get integrated together?

- How do I differentiate between Vendor A and Vendor B?

- How can I avoid getting locked in to something that does not perform?

Analytics is key to answering the first two questions, and this chapter focuses primarily on the third question about financing. The other questions will be touched on briefly here; however, they should be much easier to answer after finishing this book. It is the author's hope that information provided here will make the reader better prepared than most in the industry to answer the questions above and more. These questions are just a few of the considerations important to the stakeholders responsible for finance, operations, energy procurement, labor negotiations, risk and life safety, contract management, and other departments in facilities that influence and drive the decision making process.

As every facility and energy manager knows, without an adequate operating budget, it isn't possible to maintain buildings and equipment. Hand-in-hand with the operating budget is capital, which has been harder to assemble in recent years' economic climate, but must be available to renew and upgrade infrastructure as well as invest in analytics. In this complex environment, building owners are constrained, but open to proven business models that can have positive impacts on both budgets. Using traditional channels, the justification and approval process to get funding is a time consuming and multi-hoop endeavor even in the best of times. Elevating a typical building to "high performance" is an even more complex undertaking that requires sophisticated interoperability between dissimilar equipment and systems.

As discussed in prior chapters, access to open or standard based systems in new construction, including those who incorporate BACnet™, is not always as difficult as dealing with retrofits. In either case, the first dilemma faced by owners is choosing when to upgrade a legacy BAS system, and other building technology, and move to high-performance technologies to replace equipment. Directly related to this dilemma regarding energy and analytics is the question of what data are available from the legacy system, and would more robust data be available after an upgrade? Of course these challenges go hand-in-hand with acquisition of the analytics technology itself. The economics that support this decision process are not as clear, or well defined, as the simple ROI we are all used to—efficiencies and avoided costs after project installation equals savings that pay for the project.

The concepts of 21st century energy/electricity management and access to further smart grid capabilities, introduce new issues to consider in justifying expenditures for operational and capital improvement. When renewable energy and distributed generation resources become part of the project equation, the complexities around financial modeling increase exponentially. There have been dramatic decreases in technology cost for solar photovoltaic systems in recent years, but there is also a challenge in finding vendors that can provide these diverse offerings. If one vendor cannot provide overarching service for energy efficiency, renewable energy, DG and analytics technologies, then multiple vendors must be hired. This can add to the contractual challenges of such projects, because a new general contractor (GC) or a design build integrator must step up and take full responsibility for the project, or the manager may have to deal with a fragmented project. The price and cost of aligning the vendors can quickly outweigh the benefits. In spite of these challenges there is good news.

The good news is that a growing number of companies are providing comprehensive services. These services offer end-to-end solutions now, and are expanding to include software, hardware and web services applications to provide more solutions to the customer. These solutions consider the full energy chain from supply-side strategies selling energy, to demand-side management efficiency and capital asset management. They are beginning to also offer energy and analytics as part of the bundle. With all of these options, the questions that arise are, who is the prime vendor, what technologies do we use, how does this all fit together, and how are we going to pay for it?

In terms of a business model focusing on implementation of efficiency, renewable energy resources, DG and financing vehicles for capital upgrade projects, the energy service company (ESCO) model is perhaps the best example of a comprehensive service provider. Yet ESCOs tend to stay with status-quo measures, so they may not always be the best technology providers. The ESCO financing model however has seen a great resurgence among building owners. This is particularly true for owners who have scarce capital and/or marginal credit ratings. That said, energy services is a broad topic and includes a wide range of offerings. From a financing perspective, there are also a host of alternative financing approaches that will be discussed briefly. What is particularly desirable about the model is that it entails a combination of three critical elements to provide a design/build solution. Those elements are: 1) engineering self-funding energy measures, 2) financing based upon a revenue stream (typically from savings that are calculated and, in some cases, guaranteed) and 3) implementation. The engineering and implementation/installation elements are straightforward, but highly beneficial to managers because the design is done by energy and technology experts, who tend to have much greater expertise in these areas than architects. During this time of continuing stress on capital project funding, energy services is seeing an evolution in its financing options, beyond the "plain vanilla" versions, increasingly towards what is becoming known as ESCO 2.0.

ESCO 2.0 is a variation on the proven theme that has been called performance contracting over the past three decades. Traditionally ESCOs have brought a financing partner to the table who would structure a deal that includes construction financing and long-term debt service. The construction financing is typically handled by the same financier who provides debt service. The approach is to have the customer "close" the finance agreement, and then have the financier capitalize interest over the first six months, or install duration. Payments then begin on a certain date, so the ESCO must complete the project by that time. The actual finance vehicle may vary, but one common approach is to use equipment financing in the form of a lease purchase. Many customers buy down the first cost with cash from their own budgets, or a number of other sources, which is a good segue to ESCO 2.0. The "2.0" idea comes from Web 2.0, which according to Wikipedia "is a loosely defined intersection of web application features that facilitate participatory information sharing, interoperability, user-centered design, and collaboration on the World Wide scale." With ESCO 2.0, access to the web is critical, but the "application features" in this case are designed to access new revenue streams and create larger, more exciting capital projects. So the ESCO 2.0 phenomenon seems to be upon us, and forward looking companies have embraced it by incorporating smart grid and demand response strategies to tap into positive cash flow revenue stream that is creating a 2.0 self-funding matrix. Energy service companies or "ESCOs" are those organizations that specialize in delivering this model, but of course all ESCOs are not created equal, and it is becoming much more common for other companies to offer solutions with this same model.

This chapter is intended to reacquaint the reader with the complex topic of energy services, and to cover some of the evolution that has occurred in recent years, specifically because there is such a great opportunity to use this model to acquire energy and analytics. If the reader has not kept abreast of those changes, it is time to do so, because it could make the difference between go and no-go on your next project. This chapter will revisit the history of energy services or performance contracting, address current ESCO Best Practices and describe the Next Gen ESCO 2.0 models that are a best-kept secret.

## HISTORY

Energy services may be one of the most exciting concepts to hit the energy and buildings business over the last several decades. Yet this business model, and the industry that it spawned, has experienced many transitions over that time, and recent studies of performance contracting, combined with general observations of trends, indicate that many projects are doomed to languish without it. Equally important, energy services is poised for a next wave that will be characterized by broader forms of both technology applied, i.e. energy and analytics, and financing approaches used. Integral to these new elements will be traditional efficiency coupled with demand response using the OpenADR Standard to leverage more smart grid functionality, which will contribute more dollars to the financing pool. Analytics will be critical to leveraging the greatest potential value here through detailed analysis of building demand and consumption profiles. Before exploring all of these developments however, it may be helpful to revisit the origins of this model. The energy services model has been around for decades. This method of implementing energy and capital projects originated in Europe after World War II, when it was called "chauffage." Chauffage means "heat" in French, but at that time the idea was to address two major challenges: the need to rebuild and renew war-ravaged capital infrastructure, and the need to control astronomical energy costs for building owners. Chauffage was a solution that provided building owners with real value, delivered by a third party that turn keyed the engineering, facility management and financing for projects. At the same time, the model provided entrepreneurs with the opportunity to convert the risk associated with building renewal into a viable business opportunity. The term win-win could be used because building owners got problems fixed, investors made money, and the workforce got jobs.

The concept of energy services or performance contracting came to the U.S. in the late 1970s under a different name "shared savings." Shared savings is a concept that has seen some resurgence recently, in the form of power purchase agreements (PPAs) for renewable energy. Uncertainty in the energy markets, impacted by dramatic changes in oil/gas supply and energy price drops in the mid-1980s, resulted in changes to the model. As a result of unpredictable price volatility, the industry decided that it would exit the business of speculating on energy futures and focus on engineering of self-funding projects. As a result, the energy services model came into being and is sometimes referred to as performance contracting. This could be where a story reads, "And the rest is history." However, that's not the case for performance contracting because new electricity programs from rebates to demand response are positioning the industry to enter a new frontier.

A key point with performance contracting is that the ultimate technology limiters have to do with energy cost and the enabling legislation in a particular state. (Readers who want to learn more about enabling legislation in specific states may start by visiting http://www.dsireusa. org and clicking on the states of interest.) Critical aspects of the legislation, such as the term (number of years) that is enabled for such agreements and the types of energy measures that are allowed, will define the technologies that can be implemented. By the same token, the local cost of energy (particularly electricity), and the presence of utility rebates are also factors.

There are a number of other websites containing a wealth of information available for additional research on this topic. Two websites that include great data on performance contracting are for the National Association of Energy Service Companies (NASECO), (http://www. naesco.org/index.htm) and the Energy Service Coalition (ESC) (http://

www.escperform.org/). NAESCO, the major ESCO trade association, has a website offering information to better understand the process, the pitfalls and answers to many other questions. NAESCO, ESC, and the Association of Energy Engineers offer web resources and sponsor seminars throughout the year. The ESC is a national nonprofit organization composed of experts from wide-ranging organizations working, at state and local levels, to increase the number of building energy upgrades completed through performance contracting. This organization also offers sample documents like request for proposals or performance contract documents. The ESC website also has list of providers and other industry professionals. An LBNL study that also has good basic information that may be downloaded at http://eetd.lbl.gov/ea/EMS/EMS_pubs.html.

## THE NEXT WAVE

The adoption of IT-related nomenclature to describe next-generation buildings represents awareness that buildings are huge consumers of both energy, particularly electricity, and information. Among many important considerations is that buildings make up as much as half of electricity consumption in the US. The Energy Information Administration (EIA) puts buildings at 17% of US energy consumption overall, but buildings alone account for 35% of electric consumption, thus the building industry is very important to electricity companies. On the flip side, EIA data also indicate that electricity represents 75% of the energy bill for commercial buildings, and presents some real impact on occupants in a number of ways. Power outages decimate productivity, and according to a 2011 report from the Galvin Electricity Initiative, electricity outages cost the U.S. economy billions of dollars per year. Power disruptions also impact the work or learning being done in buildings and detracting the facility environment, lighting, HVAC and much more.

Targeting the energy cost metric, the first ESCO wave completely focused on efficiency and third-party financing, but the next iteration is addressing new funding opportunities from the other side of the meter. This author coined the term "electricity capital" to address this new category of funding opportunities. Electricity capital has been expanding rapidly, especially in New York and California where the closing of nuclear plants is driving these programs, but around the world as well. For example, in Japan there has been a huge emphasis here, in the post-Fukushima disaster years. The availability of electricity capital has been created because electric utilities across the United States, and around the world, are creating all sorts of mechanisms to promote implementation of buildings energy projects. There are a host of drivers including renewable portfolio standards, commission mandates, and forecasts that predict electricity demand to grow as economic recovery unfolds. These programs could grow even more as a result of the Environmental Protection Agency ruling in June of 2014, requiring power plants to reduce carbon emissions by 30% by 2030.

Among the many examples of electricity capital is that many utilities, both electric and gas, offer on-bill financing for efficiency projects including BAS. More examples were covered in Chapter 7. Quite simply, the building owner may be able to buy down the equipment cost with a rebate and then finance the balance of a BAS, analytics or other efficiency measure on the utility bill. Similarly Property Assessed Clean Energy (PACE) has been reborn as a vehicle for commercial financing of efficiency and renewable energy projects. The commercial PACE implementation varies by state, but under this approach building owners can typically finance improvements equaling up to 50% of the property value for up to 20 years, and the debt is repaid through an increase to the property tax. Of particular interest to many commercial property owners is that these improvement projects often increase the property value and net operating income, plus if the owner sells the building the new buyer takes on the debt in the form of higher real estate taxes. These programs, combined with high interest in energy and sustainability, are a boon to managers with scarce capital from traditional sources. As noted above, the financing models that are applied with any energy services agreement can remove obstacles to getting projects done.

Many ESCOs are teaming with smart grid and utility rebate experts, or are developing this knowledge in-house, to provide turn-key solutions. Of course many building automation manufacturers are already ESCOs, but more contractors are also adopting this model all the time. At the heart of all these new trends are smart building technologies, energy and analytics and other cloud-based software applications to enable the building to save more and fund projects. These technologies also make it possible to visualize building conditions and metrics at all times to validate the performance.

Electricity capital also includes programs like demand response

(DR) that pay customers to respond to curtailment events through deployment of strategies in the BAS to shed load, or shut down electrical equipment. OpenADR™ is the standard developed at Lawrence Berkley Labs, which many are aware of, but exhibitors at far ranging events including AHR Expo, Niagara Summit, Realcomm/IBcon, and the World Energy Engineering Congress are touting their ability to integrate OpenADR into BAS just like BACnet™ or LON™. Some manufacturers provide pre-programmed demand response algorithms, working with customers to deploy DR creating a revenue stream to fund a BAS. One DR provider has completed projects with multi-location customers that are paying for DR enablement and a new BAS with electricity capital. Under this model the owner gets technology installed with no upfront cost and is able to finance the deployment and repay it through DR payments from the utility over time. These trends are likely to continue as coal-fired power plants are shut down.

Certainly no one disputes the ESCO 1.0 business model—efficiency investments made by building owners are "bankable" and cost effective. These investments also help utilities, but only with one of their two concerns. When considering electricity, there are two important topics; energy use and energy demand. Efficiency helps utilities with energy use, and it makes sense to optimize building energy consumption because it reduces operating cost. In addition, with the end of the recession, electric demand is again growing, as discussed under the Energy Triple Threat. At the same time, emerging electricity markets present new utility concerns, and most building owners are unaware of the financial benefits that may exist in their utility territory. Those managers with graying temples remember demand limiting, etc. from the 1980s, but that's just part of the story. It is a shock that ~25% the multi-billion dollar electric infrastructure (power plants, transmission and distribution lines, etc.) exists to keep the lights on for ~100 hours/year. That's because the monopoly electricity business model has traditionally set one price for units of power, $/kWh, no matter when it is consumed. The demand charge, $/kW, is applied to commercial bills with loads about a certain limit, but that does not begin to match wholesale electric price volatility before, during, and after periods of peak electric use. ESCO 2.0 is about tapping into fees that utilities are willing to pay the owner for developing strategies to support the utility in keeping the lights on. ESCO projects have typically generated two types of savings, "energy" and "operations and maintenance." The 2.0 model will generate a new type cash flow, which is not "savings" but "income." Demand response is an example of this income that has been widely embraced, but that is just the beginning. Utilizing the same technology that has been put in place to enable demand response, many electricity users have begun to offer capacity back to the grid at peak times… for a price. The customer is paid a price that translates to electricity income. What makes ESCO 2.0 exciting is that these income strategies combined with savings strategies can create the opportunity for larger and more exciting ESCO projects

A key point here is that the 2.0, in ESCO 2.0, is first an indicator that this is truly a next generation business model. Central to this model however, is also the deployment of next generation technologies. ESCO 2.0 will provide a mechanism to achieve smart buildings, and with the energy techno-strategies deployed by local, and cloud-based technology, such as energy and analytics. With these advanced technologies, buildings can become energy profit centers by participating in both efficiency and the electricity markets. Quite simply, ESCO 2.0 will be made up of smart buildings that use energy in clean, efficient and cost effective ways and are therefore green buildings, but more importantly they are green in the environmental sense, as well as in the economic sense. Those buildings that integrate more functionality for energy services opportunity will benefit from more robust participation. ESCO 1.0 buildings often overlooked the smart clean tech implementation, but 2.0 will start with commissioned buildings that have effectively operated BAS, electric sub-metering, dashboard technology to visualize building performance and analytics to evaluate performance over time and develop benchmarks for comparison. Buildings enabled like this can participate in demand response (DR) immediately by implementing technology to receive electronic "event notifications," via the OpenADR standard, when demand response events are called. The building then reacts to that event by modifying equipment operation to reduce an agreed upon number of kW. Programs vary in the amount of notice that customers receive, usually "day ahead," "day of" or "10-minute" notice. Programs also vary in the amount of money paid for participation but customers typically receive payments between $50 and $1,000 per kW, but it can be higher. For example, a 200 kW enrollment would net an annual payment between $10,000 and $200,000 per year. With most programs the customer gets paid whether an event is called or not, and in California the utilities will even pay to install the technology. That all sounds pretty good, but these programs are evolving and expanding. As noted above, this same smart grid/DR technology can be used to enable customers to sell back, to electric markets, electricity they agree not to use. The author has called this "day trading for energy," and it gives customers another way to leverage financial value from the electricity system. Adding this income to the arbitrage mix will create even larger projects.

At first, building owners may think this is too complicated and maybe it's not worth it. The same could have been said of many technologies that are now widely used, when they were first available. That could also be said of green building recognition systems that have become the norm. Even more compelling, customers that leverage these technologies and programs effectively, can literally spin the meter backwards, while at the same time leveraging third-party capital and making debt service with REC payments from the utility and cost reduction on energy bills while optimizing building performance and tenant comfort. The ideal scenario is when customers go farther and expand building technologies to include energy storage—thermal or battery, on-site/distributed generation, or what are now called "microgrids." The idea is to take the electric reliability challenges faced in many parts of the country, and apply entrepreneurship to turn buildings into profit centers for energy. Demand response (DR) is relatively established in many parts of the country and is paying dividends for many building owners, but the next step is to insulate the tenants from any inconvenience that might be caused by DR. This is done by leveraging automation systems to pre-cool for example, focusing strategies on shutting down nonessential equipment, and leveraging on-site generation, but numerous other options exist. Using advanced DR and OpenADR based technology, owners can also aggregate across multiple buildings and opt in or opt out based on what is going on in the building. A depiction of such a building is provided below and comes from the Electric Power Research Institutes "Green Grid" report, which may be downloaded free at: www.epri.com/abstracts/pages/productabstract.aspx?ProductID=000000000001016905. Figure 17-1 shows the myriad technologies that are being applied with buildings of all sizes.

The challenge in today's economy is finding ways to fund the technologies shown in the green grid graph, and ESCO 2.0 may well be the answer. Equally exciting is that owners can leverage the intelligence deployed in these systems to make performance repeatable, and to apply energy and analytics tools that make it possible to evaluate the project in real time, not just once a year. All of these ideas mean that ESCO 2.0

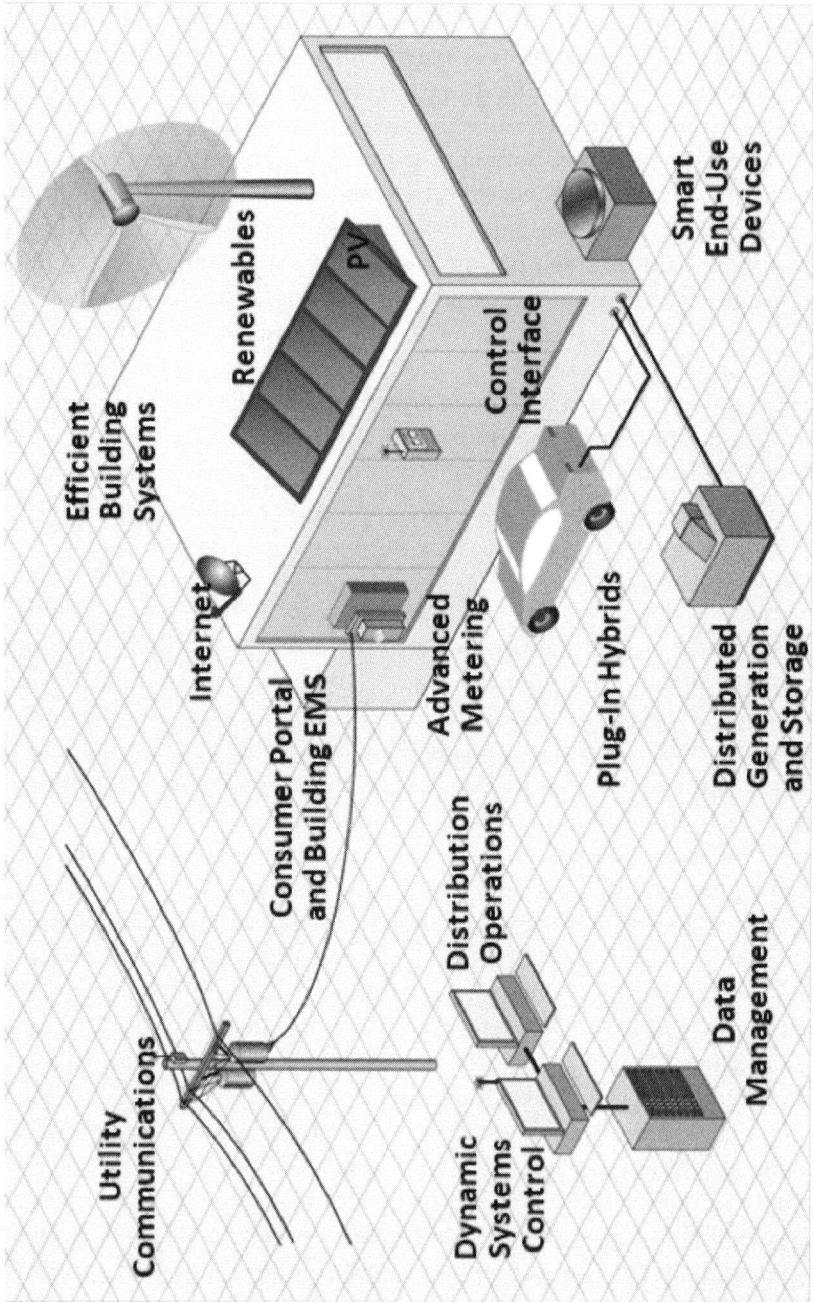

Figure 17-1

amounts to free money and no one can afford to turn that down.

The importance of the smart grid to buildings has become both real and beneficial for building managers who are getting paid to participate in demand response. Yet the smart grid, in a broader context, represents an opportunity for new building revenue streams, allowing them to become virtual power plants and energy profit centers. It is about transforming the electricity business model to unlock capital and operating cost benefits for building owners. Energy efficiency and green buildings, along with their respective benefits, have become second nature to facility professionals in the last decade, but demand response, and ultimately the smart grid, can unleash even more benefits. A great example of this fact is the launch of PEER by the United States Green Building Council, as discussed in Chapter 7. What building owners should know:

- The difference between demand response and the smart grid;

- Why it is happening;

- How buildings can benefit; and

- What it costs to play.

These questions will be answered in closing this chapter.

Figure 17-2 depicts some smart grid basics. By definition, a smart grid is an interconnected system of information and communication technologies, and electricity generation, transmission, distribution and end use technologies which will enable consumers (in this case building owners) to manage their usage and choose the most economically efficient offering, while maintaining delivery system reliability and stability enhanced by automation and environmentally optimal generation alternatives including renewable generation and energy storage. That is a bit of a mouthful, though it gets to the heart of what is underway, but why? The best way to explain this is to start with a question: What would happen if Alexander Graham Bell and Thomas Edison came back to life tomorrow and observed the industries they were instrumental in creating? If Bell were handed an iPhone™ and asked to make a call, he would not know how to do it. Edison on the other hand would be able to explain, in fairly technical detail, how every aspect of today's electric system works, even though Tesla promoted what became today's alternating current (AC)-based grid. It has not changed in ~100 years!

Bob Galvin, former chairman of Motorola and founder of the Galvin Electricity Initiative became very focused on this topic late in his

**Figure 17-2**

life. Mr. Galvin was instrumental in starting the cell phone industry, and he compared electricity in 2010 to telecom in the early 1980s—a monopoly business model, pent up need for innovation, and no way to unleash entrepreneurial business models. Speaking of business models, demand response (DR) and the OpenADR standard have proven to be killer aps in the buildings space. The term killer AP may be overused, but getting paid to implement a control strategy is pretty exciting. The OpenADR standard was developed at Lawrence Berkley National Laboratories, and it is in the vanguard of initial Smart Grid Standards to be mandated by the Federal Energy Regulatory Commission. More exciting, OpenADR is integral to demand response programs across the country, and as noted, utilities in states like California will even pay building owners to conduct a DR audit and then install the DR-enabling technology. In California, OpenADR is required for those projects. After technology is installed, the utility will continue to pay the customer every month to execute the strategies that are put in place... if needed. That is DR and OpenADR, but why smart grid? Because the electric industry is dangerously outdated and, equally alarming, the grid is increasingly unreliable. That is why power outages cost US businesses billions of dollars per year. August 14, 2003, was the day that brought this issue to the attention of congress and the American public. On that day the entire US eastern seaboard went dark from a massive power outage. This was not an isolated event; outages continue to occur regularly—thankfully, not on that scale— but with significant impacts. So aging infrastructure and an antiquated business model, along with escalating demand for electricity by our digital economy, all contribute to the reasons why the smart grid has been so important. If it could be said that there was anything good about the economic downturn, it was that electric providers got a reprieve from load growth, since electric use was down due to businesses closing, etc. But demand for electricity is growing again. A few years before the downturn, the Department of Energy was projecting a 40% increase in electric demand over 20 years, and that it would cost ~$1 trillion to build new infrastructure to keep the lights on, and those investments are still needed. Utilities and state commissions will be hard pressed to justify rate increases to fund needed investments in electric infrastructure. So they will turn to energy users with new electricity programs. Demand response is one key way to hedge against the speed at which new infrastructure must be built. Regardless, the building owner is more concerned with creative ways to meet operations and capital needs, and ESCO 2.0 is a great way to support that effort.

Again the term win-win could be applied to ESCO 2.0 because it will help renew building infrastructure and enhance electricity resilience, while hedging the need to build new grid systems. ESCO 2.0 will also support transition from fossil fuels as well as cost savings from efficiency. In closing, the term "ESCO 2.0" could be replaced with "creative financing," "electricity capital," etc. The term is not important, rather the recognition that there are a host of financing alternatives that will benefit building managers.

# Chapter 18

# Analytics for
# Measurement and Verification

There are a tremendous number of applications for energy and analytics. Many such uses have been covered throughout the book, but one that is worthy of some additional discussion is measurement and verification (M&V). Given the previous chapter on ESCO 2.0 and the many other types of financing that are being used in the buildings space, M&V is critical. Of course M&V is equally as critical when building owners self-finance energy efficiency, renewable energy and other capital investments as well. The goal of this chapter is to introduce how analytics is a game changer for M&V, but equally important how the practice of M&V could morph to incorporate commissioning and become "continuous improvement for energy management." Throughout this book, there has been extensive discussion of fault detection and diagnostics (FD&D), and a host of other analytics strategies that produce cost savings, while also providing operational benefits to improve the mission of organization inside the buildings themselves. M&V represents another potential application for analytics and is a significant topic in itself. The purpose of this chapter is to explore the idea of M&V using analytics and highlight three topics:

- M&V definition and discussion of how analytics can support the effort.

- Analytics as a stimulus that could cause the science of M&V to morph with the science of commissioning, providing a science that both evaluates performance of capital investments, and optimizes the overall building and its systems performance.

- M&V/optimization in 21st century electric markets

Analytics for conventional M&V is important to managers/owners who make a capital investment and want to ensure that the products or services purchased are producing the expected results. The science of conducting analysis to evaluate such investments is called measurement and verification (M&V), and there is an international protocol on these processes. This chapter cannot cover this topic in great detail, but it can make the point that analytics is the ideal technology for conducting M&V. It is the intent here to provide some general background on M&V and provide the reader with some insight as to how energy and analytics can offer far-reaching functionality that they may not have initially realized.

## MEASUREMENT AND VERIFICATION (M&V) SCIENCE

The International Performance Measurement and Verification Protocol (IPMVP) defines the science of M&V and provides an overview of best practice techniques available for verifying results of energy efficiency, water efficiency, and renewable energy projects. It may also be used by facility operators to assess and improve facility performance. The IPMVP began as an effort by the U.S. Department of Energy and Lawrence Berkley National Laboratories in 1994 to establish international consensus on methods to determine energy/water efficiency savings and thus promote third-party investment in energy efficiency projects. It was ultimately published under the name International Performance Measurement and Verification Protocol (IPMVP) in 1997, and seen several updated versions. The IPMVP organization grew and expanded its service offerings leading the board of IPMVP Inc. to change the name of the non-profit corporation to the Efficiency Valuation Organization (EVO) in 2004.

Capital investments related to energy in buildings and plants often include multiple individual projects, often called retrofits. Individual retrofit projects are often called energy conservation measures (ECMs). ECM is a term that usually describes one individual project, retrofit or activity that is completed in a building or plant, such as a lighting retrofit or replacement of an antiquated chiller with a high efficiency model. Of course the savings produced from multiple ECMs are typically cumulative, and they may include energy and operational cost reductions as well as qualitative operational factors that help improve how well the building accommodates the work being done there. In the 21st century energy marketplace that has been discussed throughout this book, the measures that are combined to form a project will likely encompass more than energy/water efficiency and renewable energy. There are a host of measures incorporated into projects today that straddle the demand and supply sides of the electricity system. Such measures include, but are not limited to, fuel saving, water efficiency, load shifting, demand response, participating in capacity markets for electricity, and energy reductions through installation or retrofit of equipment, and/or modification of operating procedures. Of course the savings typically include operational cost reductions as well as energy. Therefore EVO recommends that the design of a measurement and verification (M&V) plan must consider approaches that best align with: i) project costs and savings magnitude, ii) technology-specific requirements, and iii) risk allocation between buyer and seller, i.e., which party is responsible for installed equipment performance and which party is responsible for achieving long-term energy savings.

EVO has received extensive user feedback over the years and each updated version has sought to provide greater internal consistency, more precise definition of M&V options, and treatment of additional issues. The specific application of IPMVP is to measure and verify performance of projects improving energy or water efficiency in buildings and industrial plants. The current version of IPMVP is divided into three separate volumes: Volume I Concepts and Options for Determining Savings, Volume II Indoor Environmental Quality (IEQ) Issues, and Volume III Applications. More detail on content by volume is provide below.

- Volume I—Concepts and Options for Determining Savings
  — Defines basic M&V terminology
  — General procedures to achieve reliable and cost-effective determination of savings
  — Applicable to energy or water efficiency projects in buildings and industrial plants

- Volume II—Concepts and Practices for Improving IEQ
  — Reviews IEQ issues that can be influenced by an energy efficiency project
  — Focuses on maintaining acceptable indoor conditions under an energy efficiency project

- Volume III—Specific Applications
  — M&V of Renewable Energy Systems
  — M&V in New Construction

— Emissions Reduction from Energy Efficiency and Renewable
  Energy Projects

Verification of savings is then done relative to the M&V Plan for the
project. The IPMVP also makes provisions for definitions to be customized for each project, with the help of other resources.

In the context of energy and analytics, what is most important to
consider is the IPMVP option that is selected by the manager/owner.
Table 18-1 summarizes the M&V options that may be considered.

Most energy managers believe that the utility revenue meter is
the great equalizer in determining whether energy savings have been
achieved. This is particularly true when law or customer circumstances
require that savings are guaranteed (in energy units or dollars), and that
the financial value of savings provides the necessary cash flow from
savings that is necessary to meet debt payments. This balance of savings

**Table 18-1. M&V Options**

| M&V Option | Measurement Requirements | Basis of Savings Calculations |
|---|---|---|
| | | |
| Option A – Partially Measured Retrofit Isolation | Combination of measured & stipulated factors. Spot or short-term measure-ments at component or system level. Stipulated factor is supported by historical or manufacturer's data. | Engineering calculations, component or system models |
| Option B – Retrofit Isolation | Spot or short-term measurements at component or system level for constant loads. Based on continuous measurements taken at the component or system level for variable loads. | Engineering calculations, component or system models |
| Option C – Whole Facility | Based on long-term whole-building utility meter, facility level, or sub-metered data. | Regression analysis of utility billing meter data. |

vs. debt service can be precarious and of course the vendors providing services cannot control utility rates. So if savings must be at or below debt service, vendors will typically base projects on 80% to 85% of calculated savings. This means that the project is designed to hedge savings and make provisions for utility rate increases, severe weather years, some variation in occupancy, etc. Bear in mind that the use of a building is in continual flux; changes in a host of operating factors, including occupancy, must be evaluated over time to determine whether they have impacted savings. This can be more complex when dealing with campus complexes, but in those cases utility consumption baselines must be developed on a building-by-building basis to form the basis for energy savings, and M&V must be conducted on each building. A final note on IPMVP and Option C is that it assumes regression analysis on bills from the utility and based upon readings from the revenue meter. The promise of energy and analytics for M&V is twofold: 1) the revenue meter data may be complemented by interval meters for demand and consumption, and 2) this analysis can be completed electronically and available on at least a monthly basis. M&V is expensive, it is highly labor intensive and generally must be done by energy engineers, so it is usually only done on an annual basis. If a project goes off the rails early in the year, and you don't find out about a problem until months later, the outcome could be catastrophic. Coupling energy and analytics with the human expertise of certified energy managers could be a significant advancement for M&V.


ENERGY AND ANALYTICS

Again there is a data quality benefit and a timing benefit offered to M&V by analytics. These benefits can be significant to energy managers in their quest to optimize performance, control cost, and find new ways to leverage greater value from that asset. Analytics is exciting because it is a technology-rich discipline which requires professionals who understand every aspect of building operations along with the complex technology deployed. As discussed, these building technologies include metering, building automation systems (BAS), dashboards, middleware, web services/energy services tools, and these are all sources of data for analytics. Of course content knowledge is critical to analytics success; this is why M&V is normally conducted by engineers. Even with more robust comparative data from direct data access to the revenue meter or sophisticated sub-metering, it takes a trained eye to pick up some anomalies. The analytics tools can be trained to identify and report on many types of anomalies, but the ultimate question remains to be answered. What caused the anomaly and how can it be corrected? Doing the research and analysis to answer these questions requires the content expert.

Energy and analytics, as a market and a technology, are in a state of rapid growth, and enhancements are occurring at a high rate of speed. One exciting thing about this technology is that it can is being enhanced on an ongoing basis. For example, if an energy meter shows a high consumption or electric demand reading for January compared to the baseline and past months, including past Januarys, that is an anomaly. Initially the analytics tool may just flag the anomaly, but enhancements would allow the system to access data on occupancy, lighting and HVAC runtime, ancillary loads such as elevators or parking lot lights and a host of other factors that could explain the anomaly. For any particular building type, the data evaluated can be tailored to the application, and then the analytics tool can be upgraded to incorporate algorithms that look for anomalies. Coppertree analytics, a leader in this market, recently reported an excellent example of such an algorithm. In a customer's facility, the tool identified a high electric meter reading and executed a series of analyses, which identified that the parking lot lights in a 200,000-sq.-ft. parking structure were on 24/7. The tool identified the problem and the cause, so that the building owner could correct it by returning the lighting to automated control.

On a broader basis, energy and analytics can be linked with continuous commissioning for HVAC and other systems for optimization, as part of the fault, diagnostics and detection application identified in Chapter 1. The intent of this strategy would be to ensure that the facility is operating as expected by the owner and tenants. Accomplishing that goal would produce a real "mission benefit." Mission benefit means that the space is providing the expected environment for the work being done by occupants, a very important thing. For example the Building Operators and Managers Association (BOMA) has conducted surveys of tenants for decades. One of the questions on the survey is, Did you move to a new space and if so, why? Usually at the top of the list or in the top 3 for tenants who move is that the building was uncomfortable. Uncomfortable buildings are usually inefficient, so using analytics drives two benefits: cost savings (energy, O&M, etc.) and mission benefits (tenant retention, etc.). The value of optimizing building operations and energy management can be huge. This is particularly true for electricity, use and cost. Electricity is the highest quality form of energy, and is essential for a 21st century quality of life, including education, business and entertainment technology. Electric energy cost also represents one of the largest controllable operation expense in commercial buildings. Combine those facts with those of a recent report from Next 10, predicting that buildings will make up 80 percent of projected growth in energy consumption between now and 2040, and it means that the time for energy and analytics is now. Regarding analytics and commissioning, the U.S. Environmental Protection Agency (EPA) EnergyStar Program published "A Retrocomissioning Guide for Building Owners" in 2007. That guide pointed out that interval data coupled with energy analytics can help to change the game for retro-commissioning. Granular consumption data can provide a detailed view into a building's energy use. Applying energy analytics to historic data can help detect opportunities related to a building's operations and systems in minutes, without ever going onsite. Many utilities are also beginning to pay rebates for this technology and service as well.

## M&V IN 21ST CENTURY ELECTRIC MARKETS

The topic of M&V has broadened beyond verification of projects, in the new energy markets. First the process of establishing baselines, consumption profiles and benchmarks has taken on a new role for energy procurement including electricity and natural gas. Understanding how and when buildings consume energy is fundamental to negotiating energy purchasing contracts. Yet that is just the start in these new markets. The same tools that are used for M&V are often required for project verification with energy rebate programs. Most utilities require engineering analyses prior to approval of the rebate, but there are those that will require post-install M&V as well.

M&V techniques are also critical to demand response (DR) programs. Again it is critical with DR to understand consumption profiles, but also to track interval data on electricity demand. These data must then be analyzed together with an understanding of building operations and occupancy to determine whether to enroll in DR. The meter, as with IPMVP Option C, is the ultimate arbiter in DR programs. Utilities will pay enrolled customers at the agreed upon price, whether there is a DR event or not, but if an event is called and the customer does not respond there will be consequences. Consequences could mean a renegotiation of the contract, departure from the program or something else, but these issues are handled on a case-by-case basis. Therefore it is critical to have a robust M&V program. There are two final points on DR: 1) demand limiting and 2) fine tuning. Many energy managers believe that demand limiting is a better approach than DR. If it is possible to shed load on notice, during the highest period of consumption, shouldn't the manager do it every day? This is a philosophical question that will not be debated here, but may be worth some thought. The second point is DR fine tuning, which is the process of integrating DR with M&V of projects and other internal energy management programs. For example, if fluorescent light dimming was one of the strategies being used for DR and the facility is retrofitted to LED lighting, how does that impact DR commitments? This would be true if high efficiency chillers, BAS or other measure are installed as well. Generally speaking, it becomes harder to participate in DR as the building becomes more and more efficient.

The last topic under this heading relates to managing distributed generation (DG) and participation in electricity capacity markets. The basis for both of these activities is the same as described for DR, however there are some differences for each. With distributed generation, the first consideration is system type which could include solar photovoltaics, micro-turbines, combined heat and power (CHP), etc. and going forward the term will likely morph to encompass some form of electricity storage as well. Also of interest, recent EPA rules allow diesel-fired emergency generators to run up to 100 hours per year for DR. In any case effective management of DG requires an understanding of characteristics for the equipment types installed, plus interval data and consumption profiles to determine if the equipment is operated continuously or is started based on conditions. These are complex topics that will not be covered further here, but developing techniques using some form of M&V can be extremely valuable in planning and implementing these technologies.

It should be clear that the topics of M&V and analytics are, or should be, intertwined. These independent disciplines can inform development and enhancements for one another. The reader is encouraged to follow developments with the EVO and the IPMVP as the science of energy and analytics evolves. Knowledge in both of these domains will undoubtedly prove very useful.

# Part III—Analytics Success Stories

*James M. Lee, CEO Cimetrics, Inc.*

Turning data into value—that has long been the promise of analytics technology. Although applying analytics to building technology is only about 15 years old, it has already reached a level of maturity where it is beginning to produce striking results. This section contains a collected cross section of examples of analytics applied to building systems. These case studies demonstrate the technology, business process and value creation that analytics applications are bringing to the field.

In the beginning, analytics applied to building systems focused primarily on energy efficiency. Over time, people realized that energy is only one of many value propositions offered by analytics, and the technology developed to include ongoing commissioning, predictive maintenance, contractual compliance (vendor management), regulatory compliance, sustainability and comfort.

As the applications for analytics expanded, it became clear that applying this technology to building systems is a classically disruptive technology, radically reshaping competition and strategy in both the construction and operation of buildings. Power is shifting from the traditional construction and operation contractors to building owners. The information allows both transparency and accountability in an industry which has traditionally been opaque. Building operators now have key performance indicators about the health of their buildings. This new technology has created a win/win situation whereby the owner/operator, automation and mechanical subcontractors and commissioning providers can all profit while improving energy efficiency, comfort and sustainable performance of the building. Owners have seen approximately a 20% energy efficiency improvement, while traditional contractors profit from making repairs and upgrades to the existing systems.

The case studies in this section demonstrate these concepts and practices, showing the potential of analytics to change the face of building construction and operation.

# Chapter 19

# 88 Acres—
# The Microsoft Energy and Analytics Success Story

This book's author would like to thank Microsoft Corporation and Darrell Smith, Microsoft's director of facilities and energy, for allowing this document to be reprinted as a chapter here. The 88 Acres story was written by Jennifer Warnick and has been published on the internet and elsewhere, highlighting this exciting project. It has also been the topic of numerous articles, speeches and presentations demonstrating the exciting advances that the Microsoft team has made with energy, analytics and buildings. Hearty thanks are due to Jim Young and Howard Berger of Realcomm/IBcon for introducing McGowan to Darrell Smith and providing a venue to learn about the exploits achieved by the Microsoft team. As with other success stories, readers are encouraged to pursue outside research to identify recent developments.

---

A small, covert team of engineers at Microsoft cast aside suggestions that the company spend US$60 million to turn its 500-acre headquarters into a smart campus to achieve energy savings and other efficiency gains. Instead, applying an "Internet of Things meets Big Data" approach, the team invented a data-driven software solution that is slashing the cost of operating the campus' 125 buildings. The software, which is saving Microsoft millions of dollars, has been so successful that the company, and its partners, are now helping building managers across the world deploy the same solution. With commercial buildings consuming an estimated 40 percent of the world's total energy, the potential is huge.

THE VISIONARY

"Give me a little data and I'll tell you a little. Give me a lot of data and I'll save the world."
—*Darrell Smith, Director of Facilities and Energy, Microsoft*

"This is my office," says the sticker on Darrell Smith's laptop, and it is. With his "office" tucked under his arm, Microsoft's director of facilities and energy is constantly shuttling between meetings all over the company's 500-acre, wooded campus in Redmond, Washington.

But Smith always returns to one unique place. The Redmond Operations Center (often called "the ROC") is located in a drab, nondescript office park. Inside is something unique—a new state-of-the-art "brain" that is transforming Microsoft's 125-building, 41,664-employee headquarters into one of the smartest corporate campuses in the world.

Smith and his team have been working for more than three years to unify an incongruent network of sensors from different eras (think several decades of different sensor technology and dozens of manufacturers). The software that he and his team built strings together thousands of building sensors that track things like heaters, air conditioners, fans, and lights—harvesting billions of data points per week. That data has given the team deep insights, enabled better diagnostics, and has allowed for far more intelligent decision making. A test run of the program in 13 Microsoft buildings has provided staggering results—not only has Microsoft saved energy and millions in maintenance and utility costs, but the company now is hyper-aware of the way its buildings perform.

It's no small thing—whether a damper is stuck in Building 75 or a valve is leaky in Studio H—that engineers can now detect (and often fix with a few clicks) even the tiniest issues from their high-tech dashboard at their desks in the ROC rather than having to jump into a truck to go find and fix the problem in person.

If the facility management world were Saturday morning cartoons, Smith and his team have effectively flipped the channel from "The Flintstones" to "The Jetsons." Instead of using stone-age rocks and hammers to keep out the cold, Smith's team invented a solution that relies on data to find and fix problems instantly and remotely. "Give me a little data and I'll tell you a little," he says. "Give me a lot of data and I'll save the world."

Smith joined Microsoft in December of 2008. His previous work managing data centers for Cisco had given him big ideas about how buildings could be smarter and more efficient, but until he came to Microsoft he lacked the technical resources to bring them to life. What he found at Microsoft was support for these ideas on all sides—from his boss to a handful of savvy facilities engineers. They all knew buildings could be smarter, and together they were going to find a way to make it so.

Smith has a finger-tapping restlessness that prevents him from sitting through an entire movie. His intensity comes paired with the enthusiastic, genial demeanor of a favorite bartender or a softball buddy (and indeed, he does play first base for a company softball team, the Microsoft Misfits).

Ever punctual and an early riser, Smith lives near Microsoft headquarters and has taken to spending a few quiet hours at his desk on Sundays.

"I call it my den because I live a mile away. I come here, I make coffee, I have the building to myself," Smith says.

His family and the people who know him best understand. Smart buildings are his passion, and everything in his life has been moving toward finding ways for companies the world over to get smarter about managing their buildings (which will help them save money and reduce their energy use).

"Smart buildings will become smart cities," Smith says. "And smart cities will change everything."


## 88 ACRES IN A ONE-STOPLIGHT TOWN

Today Microsoft may have one of the smartest corporate campuses in the world, but in 1986, its headquarters was still a grass- and forest-covered 88-acre plot of land in Redmond, a sleepy, one-stoplight suburb of Seattle.

Back then, there wasn't even a store in town to buy underwear, and from city hall you'd have to walk to the grocery store for lunch because the nearest fast food restaurant was too far away, says Judd Black, who has worked for the Redmond planning department for 26 years.

"Redmond and Microsoft, we've grown up together, and we've learned from each other quite a bit," Black says. "We've both worked to

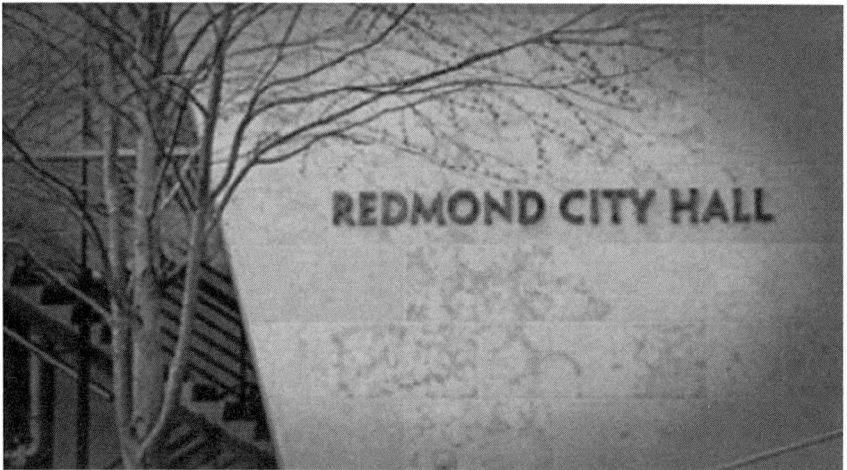

**Figure 19-1**

create a great place for people to work and live."

The 88 acres of land Microsoft chose for its headquarters (the code name for the project was just that: 88 Acres) was originally supposed to be a shopping center, but that plan was bagged during hard economic times. Microsoft snapped up the land and quickly constructed its first office complex—four star-shaped buildings surrounding Lake Bill (a large pond affectionately named by employees for founder and then-CEO Bill Gates).

Business was booming, and construction on campus followed—so quickly that Microsoft initially didn't have deeply defined construction standards. To meet demand, the company had to work with a variety of contractors and construction schedules, so consequently Microsoft's 125 buildings were constructed in a variety of styles and configurations. By the time Microsoft instituted comprehensive building standards in the 1990s, a large portion of the campus was already built.

Microsoft's campus would become the size of a small city—a city within a growing city. Redmond was expanding as well, and soon a quarter of its residents were Microsoft employees.

Today the campus spans 500 acres. There's a soccer field and cricket pitch, miles of wooded walking paths—and 14.9 million square feet of office space and labs that now function as one interconnected system.

It wasn't always that way.

Until recently, Microsoft was using disparate building management systems to manage 30,000 unconnected, sensor-enabled pieces of equipment. Imagine a symphony orchestra, but with every musician playing from different sheet music. Then, imagine trying to conduct that symphony—to make sure the music was on tempo, in key, and starting and stopping as it should. Microsoft's buildings were experiencing data dissonance that would make the works of Igor Stravinsky sound like a barbershop quartet.

Smith's team was on a journey to find harmony.

When Smith, Jay Pittenger (Smith's boss), and others started exploring ways to manage buildings smartly, they realized it would cost upward of $60 million to "rip and replace" enough equipment to get those 30,000 sensors to whistle the same tune.

This would not only involve costly construction and equipment replacement, but it also would mean displacing employees and losing work while teams temporarily shut down labs. Smith and team knew there had to be a less pricey, less disruptive way to achieve data harmony, but after a whole lot of looking, they couldn't find one.

So they invented one.

Smith's team enlisted the help of three vendors in the field of commercial building data systems and created a pilot program in 13 of the buildings on Microsoft's Redmond campus. The team developed an "analytical blanket" to lie on top of the diverse systems used to manage the buildings. The blanket of software finally enabled equipment and buildings to talk to each other, and to provide a wealth of data to building managers.

"It hasn't been a bowl of cherries—my hair wasn't as gray before we started," Smith says. "The challenge with building systems is that they can create a lot of chatter from multiple systems, but there's value there if you connect and capture it. It's all about the data. If you can't get data out of the buildings, you're done."

The new tool *did* get data out of the buildings—great tidal waves of data that came cascading into the ROC, telling engineers about everything from wasteful lighting schedules to hugely inefficient (but up until then, silent and undetectable) battles being waged between air conditioners and heaters to keep temperatures pleasant.

In one building garage, exhaust fans had been mistakenly left on for a year (to the tune of $66,000 of wasted energy). Within moments of coming online, the smart buildings solution sniffed out this fault and the problem was corrected.

In another building, the software informed engineers about a pressurization issue in a chilled water system. The problem took less than five minutes to fix, resulting in $12,000 of savings each year.

Those fixes were just the beginning.

Suddenly, the symphony of sensors was not only following the conductor, its musicians were all playing the same song. As buildings came online and data poured in, it created what engineers called a "target-rich environment" for problem solving. Smith and the team soon expanded the pilot to a handful of additional buildings, and by summer's end they plan to have the whole Redmond campus online.

The team now collects 500 million data transactions every 24 hours, and the smart buildings software presents engineers with prioritized lists of misbehaving equipment. Algorithms can balance out the cost of a fix in terms of money and energy being wasted with other factors such as how much impact fixing it will have on employees who work in that building. Because of that kind of analysis, a lower-cost problem in a research lab with critical operations may rank higher priority-wise than a higher-cost fix that directly affects few. Almost half of the issues the system identifies can be corrected in under a minute, Smith says.

The change has created groundbreaking opportunities for Smith and his team.

"Our conversations have changed," Smith says. "Before, the calls we got were about buildings being too hot or too cold, or about work orders. Now we're talking about data points and building faults and energy usage. We're seeing efficiencies that we never even contemplated when we started this journey."


THE LIVING, BREATHING BUILDING

Tearle Whitson and Jonathan Grove stroll across the rooftop of Microsoft's Building 88. Washington State's Cascade Mountains.

"This is one of the perks of the job," says Whitson, taking a moment to survey the scene before climbing into the whirring, dark interior of one of the roof's large, white air handlers. As nice as they can be, now rooftop field trips like this are few and far between.

Whitson and Grove have experienced a seismic shift in their workday since helping to develop Microsoft's smart buildings tool. Two years ago the two spent a lot of time climbing over rooftops, inspecting pump rooms, and peering above ceiling tiles at variable air volume (VAV) boxes.

"I used to spend 70 percent of my time gathering and compiling data and only about 30 percent of my time doing engineering," Grove says. "Our smart buildings work serves up data for me in easily consumable formats, so now I get to spend 95 percent of my time doing engineering, which is great."

Before Microsoft's buildings leapt up the IQ curve, the duo's home was on the so-called range. They'd move from building to building, camping out in each for two weeks at a time to inspect and tune it top to bottom before moving on to the next. It would take them five years to tune up all of the buildings on campus, and then they'd start the process all over again. Their tune-ups were making the buildings run more efficiently, saving the company around $250,000 annually—but the new data gold rush will help them save six times that much.

The duo now spends most of their time at the ROC, chewing on building data. Though they're no longer camping out together tuning up Microsoft's campus the old-fashioned way, the two have maintained the comfortable rapport and geeky banter they established working in the field.

**Figure 19-2**

Facilities engineers like Whitson and Grove think of the buildings they care for as living, breathing things. Just like the human body, buildings have a wealth of indicators that things are going well—or, in some cases, not so well. Also like the human body, small ailments can lead to much larger failures, and an ounce of prevention can lead to many pounds of cure.

And now, with the new data-driven software solution that the team built, they can do an even better job of managing the health of Microsoft's buildings—they've gone from country doctors, tapping a patient's knees with a rubber hammer, to specialists with an MRI machine who can examine every layer of the knee inside and out.

Still, Whitson says it can be a tad unnerving to take a building and its network of sensors online and watch as the software immediately discovers a host of inefficiencies.

"There is a little bit of a mindset among facilities engineers. Everybody is prideful, and they take such ownership of their buildings, that it's hard for them to find out that there was a lot that they were missing. It was a humbling lesson I had to learn early on while doing this," Whitson says. "We have to get the old-school technicians out there to understand that this is going to help them. This is not to say you've been doing it wrong—you're doing fine. But you can go farther."

Both are quick to tell you that Microsoft's smart buildings solution has revolutionized the way they and their fellow engineers work.

"We had the perfect environment and people to put all the pieces together. Our solution is a little unique to Microsoft, but very applicable industry-wide," Grove says.

ENOUGH DATA TO CHANGE THE WORLD

The ROC itself looks like an air-traffic control tower—albeit without the windows.

A bevy of large, wall-mounted displays wrap around the interior of the room, and the same monitors sit atop every desk. The smart buildings tool dashboard is splashed across many of the screens, showing off a colorful collection of maps, dials, lists, and tickers.

Engineers can get big-picture information at a glance, like how many kilowatts of energy are being consumed across Microsoft headquarters at any one moment. With a few clicks they can also zoom in on one building, one floor or office in that building, or one piece of equipment.

"Let's see how City Center building is doing today," Whitson says, and within seconds he's clicking through a wealth of information about the building—the number of employees who work there, the outside air temperature, the thermostat, what time the lights come on and go off, even a list of mechanical inefficiencies the software has detected and how much each of those faults is costing the company per year.

Once Whitson sits down in front of one of the terminals, he is suddenly connected to Microsoft's buildings. Those 500 million data transactions per day can be accessed from the dashboard at his desk rather than from crawling through pump rooms or across rooftops to get data.

Now that the smart buildings software cooks up a chuck wagon of data every day, what to do with all of that tasty information?

The software identifies issues large and small, and even puts them in prioritized order according to how much the problem is costing the company. A majority of problems they can fix right from their desks, and for the rest, the engineers issue work orders (about 32,300 per business quarter).

Apart from efficiency, the surge of data has also made for some eye-popping analytics. These are mechanical engineer Trevor Sodorff's

**Figure 19-3**

specialty. "We have good people, but without good software there are limits to what you can do," Sodorff says. "Everything lives within the context of the bigger picture."

One of Sodorff's party tricks, if you will, is whipping out algorithms to detect new mechanical faults. So during meetings that wander through stretches that don't pertain directly to him? Rather than discreetly checking his email or letting his mind wander, Sodorff writes algorithms.

At one such meeting, Sodorff announces that he's just written a new algorithm for detecting when the air in a given building is being overcooled. He projects the algorithm on a screen, and then launches into a deeply technical explanation about when a discharge air pressure set point is something-something, then the air is being overcooled by something-something for a duration of 900,000 milliseconds.

"That's 15 minutes," says Grove, his fellow engineer, translating on the fly.

Later in the meeting, Grove is talking about how the smart buildings software helps the engineers measure and validate that the energy reduction they're

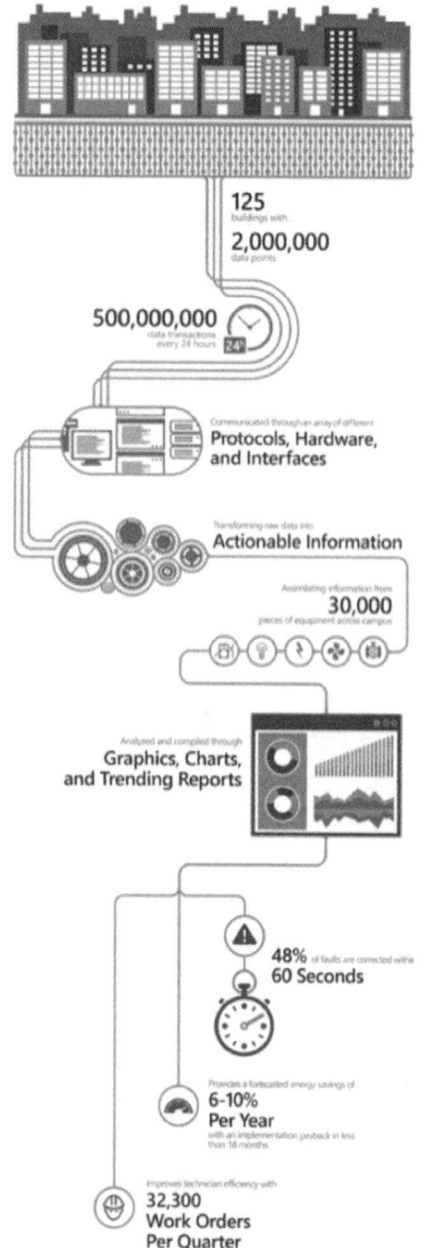

**Figure 19-4**

seeing is due to reduced consumption and not because it was 5 degrees cooler than yesterday. It's an important distinction for companies to make, especially when seeking a utility rebate.

"We may do an audit, and find we've done something that saves 200,000 kilowatt hours, which works out to … uh …" Grove says.

"Sixteen-thousand dollars," Sodorff says without missing a beat. ("I studied math for two years before I decided to be an engineer," he adds. "It's what butters my bread, if you will.")

"Never in my wildest dreams did I think I'd be presenting at the Pentagon. It was a thrill."

Darrell Smith beams at his team.

"Now you see why developing this software at this scale was a once-in-a-lifetime opportunity," Smith says. "It affords you the ability to work on some very large-scale, world-changing projects with some very smart people. It's Microsoft University."

If the smart buildings tool was developed at Microsoft University, today is graduation day. Where much of Smith's time the last few years has been spent developing the software, he now spends hours with visiting business, government and industry leaders offering enthusiastic show-and-tells. He's presented to hospitals, oil companies, automobile manufacturers, cities, and federal government agencies—even at the Pentagon and very soon, this same solution Microsoft has deployed will be available to any business.

Smith says. "It's been interesting, because I don't see myself as a salesperson. I see myself as an evangelist for the smart building industry, and what can be achieved with smarter buildings."

Office buildings, hotels, stores, schools, hospitals, malls and other such commercial buildings are responsible for up to 40 percent of the world's total energy consumption. In the U.S. alone, businesses spend about $100 billion on energy every year.

"Buildings have been built and run the same way for the last 30 to 50 years," Smith says. "This isn't a Microsoft problem, it's an industry problem."

## A SMARTER FUTURE

In one memorable scene from the 1987 movie Wall Street, Michael Douglas stands on a beach, hair slick and wearing a black bathrobe, barking orders to his protégé via a brick-sized cell phone as the waves crash behind him.

It was one of the first times a mobile phone appeared in a movie, and at the time talking on the phone on the beach demonstrated the high-tech fruits of a stock broker's wealth and power. Twenty-six years later, the scene feels dated—solely for technology reasons.

That brick with an antenna that Douglas is holding? That is the current state of commercial buildings, Smith says.

"At first, all mobile phones were bricks—basically two-way radios. Now, in just a few short years, they have advanced to become a laptop in your hand," Smith says. "Buildings are still that brick phone. We want to get buildings to where phones are."

Microsoft's campus went from bricks to brains, and Smith believes all commercial buildings can follow suit.

Jessica Granderson, a commercial building and lighting research scientist at Lawrence Berkeley National Laboratory, agrees with Smith. She says smart buildings are becoming more prevalent, and the commercial real estate industry is going through a "time of rapid change and maturation."

Granderson, whose research focuses on intelligent lighting controls and building energy performance monitoring and diagnostics, says the commercial real estate industry is reaching a major tipping point as people realize the power of capturing and analyzing data from buildings.

Granderson says it's still not as easy as it should be to get data out of today's buildings.

"There are two sides to the coin really. There's data that is out there that's not put to good use, and businesses that aren't using it. The other side of the coin is that once we do become in the habit of making use of what's there, then you quickly realize the challenges of getting what you really need and want," she says.

Another challenge is having the right people in place to analyze and interpret the data and act on it.

Granderson contributed to a white paper on Microsoft's new software, "Energy-Smart Buildings: Demonstrating how information technology can cut energy use and costs of real estate portfolios." The paper investigates and evaluates Microsoft's smart building tool, which Granderson says deserves to be held up as a best practice in the industry.

"It's one of the more sophisticated implementations that I've seen," she says. "There are a lot of lessons to be learned (in what Microsoft created)."

Probably the most important take-away from Microsoft's smart buildings breakthrough is just how much money and energy businesses can save with relatively little up-front investment, says Jim Young, CEO of the commercial real estate and information technology company Realcomm.

Young has traveled all over the globe for more than a decade, visiting and studying all manner of smart buildings and smart campuses.

"Companies, even a lot of Fortune 500 companies, have these massive real estate portfolios that they're running with sledgehammers," Young says. "What Darrell and his team can do is watch their buildings at a rate no human has before. When you start connecting to the buildings, generating data, and grouping inefficiencies by cost and priority, all of a sudden you go from the sledgehammer to running buildings with laser precision."

Smith's plan to take Microsoft's smart buildings software worldwide is to adopt the role of matchmaker. His team developed the smart buildings software, with the help of vendors, exclusively with off-the-shelf Microsoft software such as Windows Azure, SQL Server and Microsoft Office.

Smith says these partners and vendors are eager to help businesses of every size, shape, and need to take their buildings from piles of bricks to data-driven brains. Some companies need only a little push—the know-how to incorporate weather data, or energy meters, or perhaps just the right connections. Others need both hands held, procedurally and technologically. Regardless, Smith and his partners can help.

"People may be bumping their heads, but they come here and get a glimpse of the potential," he says.

Young says Smith has been talking about the possibility of making buildings smarter since the two met more than a decade ago at a Realcomm conference on the collision of commercial real estate and information technology.

"He'd say, 'I'm going to figure this out.' I'd call him every few months and say, 'How are you doing, Darrell?' and he'd say, 'I'm not ready to talk to you yet.' Then, when he finally had something to show, he was like, 'Here it is,' and it blew us away," Young says.

Though Smith and his team were quiet for the years they spent developing the software, their work has already garnered serious attention in the industry.

There's now such an interest in smart buildings in general—and Microsoft's smart buildings software in particular—that Realcomm hosted a conference on smart buildings in the Seattle area in 2013. He held it there so attendees could visit Microsoft and experience one of Smith's enthusiastic show-and-tells.

"Darrell's going to be a rock star. Well, he already is in our world," Young says. "If he's not shaking the president's hand in a year then we've all done something wrong."

Unsurprisingly, Smith is more measured about his part as the leader of a team that developed a game-changing smart buildings tool. Though his team started with little more than a notion that there had to be a better way, they used Microsoft's own software, their own background and expertise, and an unyielding determination to solve an industry-wide problem. Smith played the roles of catalyst and mortar in this process, sparking the revolution but also helping to hold his small team of "ninja innovators" tightly together.

"It's not me that's great, it's our story," Smith says. "Yes I'm passionate, but you could put someone else in my place and it would still be a great story."

**Figure 19-5**

# Chapter 20

# Smart People, Smart Grid

*Ron Rajecki, Contributing Editor, Contracting Business Magazine*

This book's author would like to thank *Contracting Business Magazine*, and Ron Rajecki, Contributing Editor, who authored it, for allowing this article from October of 2009 to be reprinted as a chapter here. The project described here was an early implementation of technology types and approaches that have been discussed throughout this book. The reader will note how the article points out many of the challenges and obstacles that have been faced by owners of complex building systems. Equally exciting, it should be evident that the advances underway in the industry, as outlined in this book, will simplify these types of project solutions going forward with more standardized products and web services. The University of New Mexico has been an early adopter of new technology. In the interest of full disclosure, the book's author is former CEO of Energy Control, Inc., so that distinction has been made as appropriate. Further, the article appears here as written by Mr. Rajecki and published in the magazine in 2009. As a result, some of the facts, figures and government references may no longer be accurate. Particularly for future projections, the reader is encouraged to consult updated information.

---

An intelligent energy system at the University of New Mexico proves electric systems can be designed to be more reliable and efficient. Should there ever come a time when cities of the future are powered by "smart grids," projects such as one undertaken by Energy Control, Inc. (ECI), at the University of New Mexico (UNM), will be looked upon as one that led the way to that future. For the present time, we gladly award this project with a 2009 ContractingBusiness.com Design/Build award.

The University of New Mexico smart grid leverages smart buildings, smart meters, thermal storage, renewable energy, and distributed generation. For those not familiar with the concept of the smart grid, Jack McGowan, CEM, former CEO of ECI, is happy to provide a brief primer. After all, he was a founding member of the U.S. Department of Energy GridWise Architecture Council, and served as its chairman for two years.

"August 14, 2003, when the eastern seaboard of North America went dark, was a significant date in electric industry history that captured the attention of both the media and Capitol Hill," McGowan says. "It quickly became evident however, that the blackout was a symptom of a much larger problem, which many have begun calling the 'energy perfect storm.' This problem begins with an aging electric infrastructure that has not had major technological advancement in a nearly a half century. It is exacerbated by a U.S. Department of Energy (DOE)-projected 40% increase in electric demand over the next 20 years that will cost utilities a projected $3 trillion. That includes the impact of cap and trade, which is part of the third element of this perfect storm: climate change."

Thus was born GridWise™, an initiative created under the DOE Office of Electricity and Energy Reliability. The focus of GridWise is on stimulating the development and adoption of an intelligent energy system, or smart grid, to make the U.S. electric system more reliable and efficient.

UNM was an ideal site for a GridWise project because of ongoing technology investments the university had made. UNM had built a mechanical engineering building in the 1980s equipped with solar thermal and thermal storage, but the systems had fallen into disrepair and were not operating. In fact, Andrea Mammoli, Ph.D., professor of engineering at UNM, stepped up to lead a team of academics and facility engineers to work with ECI on the design and construction of this project. The university had also invested in a smart meter system, a district heating and cooling plant with power generation capability of eight megawatts of combined heat and power, and numerous automation and integration technology projects over several decades.

The advent of smart grid, according to McGowan, presented an opportunity for a truly cutting-edge Design/Build energy project, and the ultimate Green Building system. Using this rationale, ECI was successful in acquiring around $600,000 in pre-stimulus DOE funding, partially matched by UNM and the state of New Mexico. "This was a perfect example of the type of design/build shovel-ready projects sought under the American Reinvestment and Recovery Act," McGowan says.

Figure 20-1

This design/build energy technology project leverages existing technologies that have been implemented at UNM, with new technology and Internet-based web services. "This unique Design / Build project leveraged wide ranging relationships and skills," according to McGowan. "ECI has had a strong presence in building automation and in the smart grid market since its inception. We recognized the connection between smart buildings and smart grid, and saw how this market offered explosive opportunity for design build solutions that blend automation, system integration, mechanical/renewable energy systems, and efficiency."

## AN INTUITIVE TOOL TO UNDERSTAND ENERGY USE

This Design/Build energy technology project leverages existing technologies that have been implemented at UNM, with new technology and Internet-based web services. The project touches many campus buildings with automation, metering and integration. In addition, an "energy business intelligence tool" combining energy and analytics with dashboard technology provides full information for more than 100 buildings. Energy data from the mechanical engineering building, shown here, is reported in real-time through an integration using Tridium Java Application Control Engine (JACE) and Delta Controls technology. Energy data from 84 other buildings is reported in less than 20-minute intervals and can be displayed in key performance indicators such as energy consumption or cost per square foot. The introductory graphic in this chapter was used by the DOE in a poster session to depict this project.

"This system provides an intuitive management tool for building owners to understand the impact of energy on their business and to provide for real-time management of building performance," McGowan says. "The system also integrates multiple BACnet™ and legacy automation systems to provide seamless interface to building operations. This smart campus can use this tool to manage energy use, energy cost, and its carbon footprint." ECI worked with an Analytics and Dashboard vendor and Delta controls to deploy a seamless BACnet integration between the BACnet systems in more than a dozen campus buildings and the Dashboard. A similar integration was executed between Tridium JACE and the Dashboard to access Btu energy data from systems that

**Figure 20-2**

pick up this data from industrial controls on HVAC.

The project deploys automated demand response (ADR). ADR is a technology that integrates utility systems with home and building control and energy information systems. Technical development and software programming at each customer's site may include a smart thermostat or a building automation system programmed to shut down equipment and reduce electrical demand if it receives a signal to do so. A signal could also cause the start-up of an emergency generator or a CHP system to generate power on campus when the utility grid can't keep up with demand.

"The UNM project was designed to show how this level of performance data could be combined with automation and energy technology to provide a new generation of smart green building. We believe that smart buildings use energy in a clean and efficient way, to become green buildings," McGowan says. The site's "ultimate" green building is—not surprisingly—the mechanical engineering building. As part of this project, ECI's team rebuilt the solar thermal system with vacuum tube collectors and 400,000 gallons of thermal storage in the building. The project completely upgraded the system, added a 200-ton absorption chiller, and connected the building to the campus chilled water loop. As a result of the work in just this one building, the project can actually take the entire building "off the grid" for cooling, with the exception of some fan and pump loads. Centennial Engineering, the newest building on campus, was designed for optimum energy performance including a connection to the thermal storage loop from the mechanical energy building.

"The total campus project shows the next generation of smart building/smart grid strategies for energy management and load management, to support the power grid. It provides a great example of optimized Design/Build delivery," McGowan says.

## INSPIRING FUTURE GENERATIONS

Mammoli says working with McGowan and the team from ECI has helped him see the real-world, practical applications of the theories that underlie the smart grid. And while no one really knows exactly what a city-wide smart grid is going to look like, or how it will ultimately function, putting one on a college campus is the right thing to do. "The campus is like a small city, so it's very exciting to look at the potential of smart grids, and to think that someday we'll say this is one of the places where it all started," Mammoli says. Mammoli is not the only one excited by the technology. "One of the things I really enjoyed about working with ECI is that doing practical things like this on buildings and energy systems really fires up students," he says.

"There are a lot of students that are really interested in this topic, and I think the interaction with ECI has been instrumental in sparking that interest. This is where the people who are going to be doing this type of work in the future are going to see it."

# Chapter 21

# Envision Charlotte:
# Energy Big Data at Community Scale

*Contributed by Tom Shircliff*
*Chair, Envision Charlotte and Principal, Intelligent Buildings, LLC*

## AUTHOR'S NOTE

*An initial premise for this book has been that the technology described presents huge potential benefits for energy efficiency, but more importantly for the owners and occupants of buildings. Envision Charlotte is a great example that this is true, but more importantly that those benefits extend to the larger community that is shared by those occupants and buildings. This is one of the most exciting stories shared in this book, and I want to sincerely thank Tom Shircliff for writing such a masterful account of this city's success. Tom's work with Envision Charlotte and Intelligent Buildings, LLC, is exemplary.*

## INTRODUCTION

Envision Charlotte is the ultimate big data energy and analytics project that leverages millions of square feet of near real time energy usage pumped through wired and wireless networks into the cloud and back out to data bases, web sites, social media and digital signage.

However, before Envision Charlotte was associated with big data or analytics, it was considered an economic development, "smart grid" and community alignment effort. In this case the community was uptown Charlotte, NC, which is a well defined urban business district comprised of 65 office buildings totaling 25 million square feet, dense residential towers, retail and other use-types with a weekday population of approximately 75,000.

The big idea was simple: connect and measure energy as a community with a single number and work together leveraging behavioral science to drive down usage. This concept of connect, measure and move the needle was so understandable and powerful it was accepted as a Clinton Global Initiative that subsequently exceeded it's three-year goals for energy reduction according to the North Carolina utilities commission[1].

The effort was borne from a long-term economic development planning process that sought a way to use energy and sustainability to drive economic activity as well as attract and keep talent. It would demonstrate a progressive, cooperative and innovative spirit in the city that is creating excitement and a new dimension to branding.

This revolutionary approach is changing the way we look at urban business districts and adds great context to how big data and analytics can facilitate and measure energy, sustainability, community engagement and economic development.

This has become the elusive definition of a "smart city," which is a focused, measured effort that has a combination of stakeholder alignment, technology and a unifying purpose.


A COMMUNITY MODEL

Along with unprecedented sustainability, economic and quality-of-life improvements Envision Charlotte is creating the model for how community stakeholders can work together including government, business, real estate owners, utilities and academia as well as solution partners. While Charlotte has a history of cooperation to achieve big things, this offered a programmatic and repeatable model based on the unifying theme of energy and sustainability being leveraged for economic development.

Envision Charlotte solution partners have made substantial investments totaling over $5MM—for good business reasons—and their results and reactions speak for themselves.

*Cisco CEO John Chambers says "With Envision Charlotte we can completely transform a city... suddenly we see a model for the rest of the nation and the rest of the world"*

This community model is built to be manageable, measurable and replicable by approaching cites and regions one sub-market at a time for fast implementation and clearly measurable results. The first sub-market was the downtown urban core, which is a place that is described as "small enough to manage but big enough to matter." That means there is a nimble structure for fast-decision making, innovation and actual implementation—but with 75,000 people, 25 million square feet, multiple corporate headquarters, city and county government, professional sports venues, a transportation hub for buses and light rail, world class hotels and fine restaurants there are all the elements of any world-class, urban community.

The concept of connect, measure and move the needle was so simple and powerful there was a movement to expand beyond energy to include air, water and waste. Thus a non-profit was formed that has developed a similar water program and likewise waste and air quality programs are being formed.

The approach is very business-like requiring all efforts to "plug in" and measure whatever it is that is being measured not only in energy but in the other three pillars of air, water and waste. You can't manage what you can't measure and if you are serious about sustainability, business and economic development you have to measure like that by capturing and analyzing the data.

Envision Charlotte also engaged or communicated with leading experts from MIT, Wake Forest, University of Pennsylvania, UNC-Charlotte and sustainability, behavior and economic experts from around the country to hone the approach and results.

A BUSINESS MODEL

The base business model was an innovative approach to the existing utility incentive program. Unlike many other Smart community initiatives that implore or require private building owners to invest capital infrastructure in their buildings as part of a coordinated effort, Envision Charlotte took a different approach. Since the goal was 100% participation of the building owner community there would be no capital investment requirement from them. The only request was that they allow the Energy data from their buildings into the program and that they allow some equipment be installed in their building and the lobby. Additionally, there would be some coordination required other facility management staff. The Capital investments required for this program would come from the electric utility Duke Energy. Duke energy would get cost recovery and a return based on the avoided cost of energy generation equal to the energy that the program saved. Awareness and behavior programs have been tried all over the country with varying results and very few successes at a scale like this. The challenge has been getting accurate measurement and verification results usually because of a very wide geography or and inability to get actual data. The foundational attributes of this approach that enabled accurate measurement are limited geography, limited use type and hard, real-time data acquisition from the buildings with a combination of technologies.

**Technology and Program**

Big data is a general term that can imply many different technology components based on the application. In this case there was a powerful combination of the cloud wide-area networking, local area network wired, and wireless Smart grid technologies and analytics. Duke energy leveraged key partners including Cisco Systems Verizon wireless and intelligent buildings LLC (contributing author). Duke energy implemented various smart grid technologies including smart electrical meters for near real-time Energy Data, as well as their secure and private customer information software systems. Verizon wireless offered flexible fast wireless connectivity required to connect dozens of buildings in a dense permanent environment. Cisco Systems provided the networking gear as well as Digital media infrastructure and equipment such as large interactive touch screens and related equipment. Intelligent buildings focused on the overall architecture, program approach and real estate owner perspective.

**Differentiation**

Envision Charlotte studied and met with many of the highest quality and best-known efforts such as Austin's Pecan Street Project, The Portland Sustainability Institute, Greenworks Philadelphia, Denver Living City Block and others where a great deal was learned from their successes and challenges. Envision Charlotte gained clarity on several key differentiators in its approach including:

1.  Data: Connecting and measuring with the detail, scale and data points

2.  Focus: Laser focus of geography and use-type

3.     Business: Depth of involvement with the business community and economic development purpose

**Results**

The results from Envision Charlotte have exceeded expectations both objectively and subjectively. These are just some of the documented numbers in this growing effort:

- 98% building owner participation

- 6.2% net savings from awareness and behavior

- 8.4% net savings from awareness/behavior and ECMs[1] (two years remaining)

- Over 1,000 "energy champions" trained

- $10MM in energy operating expense savings

From the beginning it was clear that Envision Charlotte was meant to be a true community program, which required 100% or at least very high participation from building owners. Some other so-called smart community programs asked a limited number of building owners to make capital improvements and highlighted them as examples. However, this does not have broad enough impact or representation to be called "community."

Therefore, a deep understanding of the real estate culture and decision-making process is necessary joined with a program that would gain acceptance from all points of view. The formal participation rate of 98% is perhaps one of the most significant accomplishments and a beacon for other similar community efforts.

The hard numbers for both overall and awareness and behavior energy savings have been marked as a milestone for the awareness and behavior industry. There have been many types of attempts across the country and the world but few that have assembled the components to connect and create hard measure for such a large number of people and square footage.

At the heart of the community are the tens of thousands of building occupants who impact energy and also take the training and experience of Envision Charlotte with them wherever they go. The over 1,000 energy champions, many more that connect through social media, the facility mangers participating in training and town hall meetings and the various other interactions all redound to a convening power and connectivity that will have lasting affects.

Underneath all of the aforementioned numbers, perhaps the most sustaining figure is the more than $10 million dollars in real estate operating expenses that have been saved through this program. This creates financial incentive, economic attractiveness and reinvestment opportunities that will attract and keep business and talent in Charlotte.

**Reference**
1.   NC Utilities Commission filing 5/13/2014 Document number F-E-20140513-016

# Chapter 22

# Medical Center Energy and Analytics Case Study

*Jim Lee, President/CEO, Cimetrics Inc.*
*David Landman, Chief Energy Engineer, Cimetrics Inc.*

Cimetrics, Inc. is an industry leader in energy and analytics. This case study in a medical center is an excellent example of how the technology that has been discussed throughout this book can be applied to critical care environments.

## INTRODUCTION

In the past 15 years since Cimetrics introduced automated monitoring based commissioning and automatic fault detection and diagnostics to the buildings industry, there have been many innovations and new entrants into the market. Through the fog of marketing rhetoric, it has become difficult for building owners to understand what analytics are and their importance to building commissioning. This case study will introduce analytics and fault detection as applied to building operations and demonstrate how the technology and applications deliver value.

**Analytika** is the process of extracting value-adding insights from data. Insights can come from understanding what has happened in the past and from predicting what is likely to happen in the future.

Historically, manual analysis has been performed; temperatures and pressures have been monitored and occasionally, trend analysis has been applied to building automation data. What is new is the ability to gather and process much more physical data than in the past. By using modern big data approaches, which apply algorithms to system models, the effectiveness of monitoring based commissioning is greatly increased.

Cimetrics Inc. through its analytika brand provides ongoing commissioning software and services to owners and occupants of commercial, institutional and industrial buildings. Analytika collects real-time data

335

from a client's BASs and sub-meters, integrates information from multiple facilities, and applies over 1,000 proprietary algorithms to the data. Analytika identifies opportunities to reduce energy, environmental, maintenance, operations and regulatory costs; uncovers potential equipment problems; points to profitable retrofit projects; improves occupant comfort; and enhances facility operations and uptime. By correlating control system data with data from networked switchgear and meters, analytika also enables demand response programs, studying mechanical, electrical and control systems' behavior to select and optimize the sheddable load in buildings without compromising mission-critical activities. The analysis of actual building performance during demand response events over time can improve the buildings' response to such events.

After connecting to a building automation system, data relevant to the analysis are transmitted over a secure internet connection to the analytika data center. Cimetrics' energy, electrical and mechanical engineers use proprietary algorithms and software to analyze building efficiency. Analytika algorithms apply static data (equipment specs, system topography) and weather information to the operational data collected on each piece of HVAC equipment as well as the entire building's mechanical system. The algorithms have been designed based on standard energy engineering techniques (such as bin analysis) and academic research which provide an independent measurement of building system efficiencies. The purpose of the analysis is to identify, quantify and prioritize viable energy and operational cost savings opportunities for all aspects of a facility: control systems, HVAC systems, lighting, electrical distribution, metering, operational and other equipment, and scheduling procedures. Figure 22-2 shows a screen shot from the tool with one example of such energy savings calculations.

An analysis of the facility is made including investigation of electrical power usage of motors, air delivery from fans, combustion efficiency of boilers, lighting intensity, etc. The analysis also includes interviewing appropriate administrative and maintenance personnel regarding equipment usage and operating schedules. These data are used to model the facility for analysis. Definitions of building shape, size, construction, occupancy, lighting, temperatures, schedules, controls, weather locale and other details are used to create a computer model to optimize cost saving opportunities for the facility. Figure 22-3 shows an overall summary of analysis conducted and opportunities identified from the analytika dashboard.

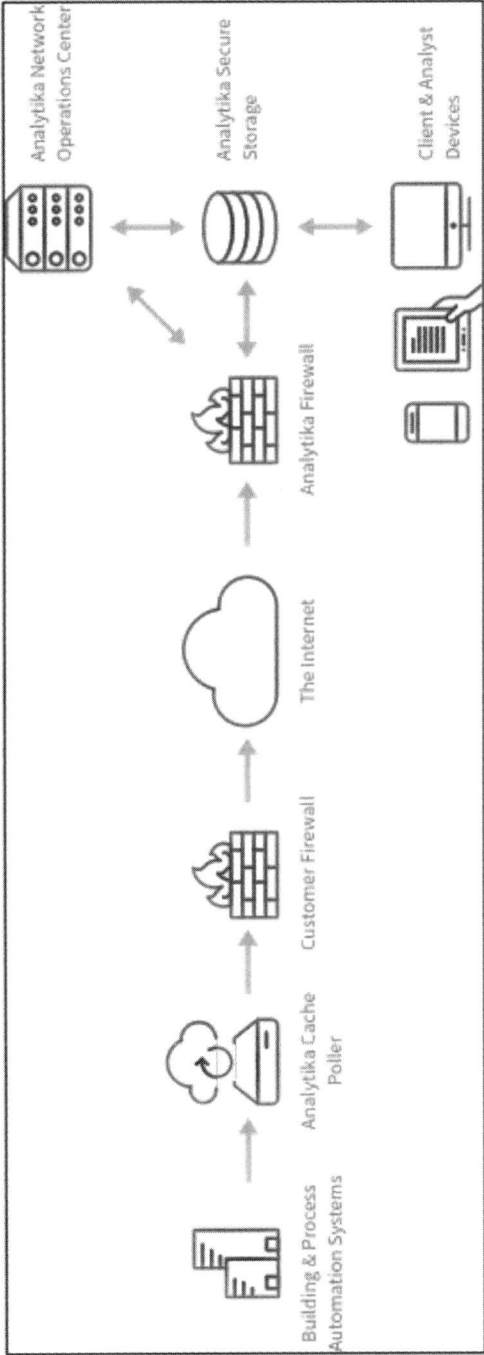

**Figure 22-1**

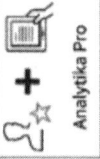

# Energy savings and impact analysis

Analytika Pro

**Energy savings calculations and assumptions:**

Potential energy savings calculations from installing a VFD, not using the bypass valve, and reducing DP and the associated assumptions are shown below.

Methodology and assumptions:

- TMY2 hours for BOS > 28 F = 7,906
- Pump = 23 BHP (from mechanical plans)
- 0.746 kW/hp
- Controls info indicates sized for 20 psi DP, not 40 psi
- Savings based on using VFD instead of DP valve and reducing the hot water loop differential pressure setpoint from 40 psi to 25 psi. Resetting the hot water loop setpoint results in lowering the pump motor speed. Pump head = 150 ft. - from mechanical plans
- Use the Pump law relationship between Power and Pressure

$$\frac{PWR\_ELEC}{PWR\_ELEC_{Full}} = \left(\frac{HEAD}{HEAD_{Full}}\right)^{\frac{3}{2}}$$

- 1 HP = 0.746 kW
- Assume 90% pump motor efficiency
- Blended electric rate of $0.115/kWh

Annual Cost Savings: **$18,359**

Annual kWh savings: **159,641**

Annual $CO_2$ Savings: **83.25 Metric-Tons**

ANALYTIKA   Confidential

**Figure 22-2**

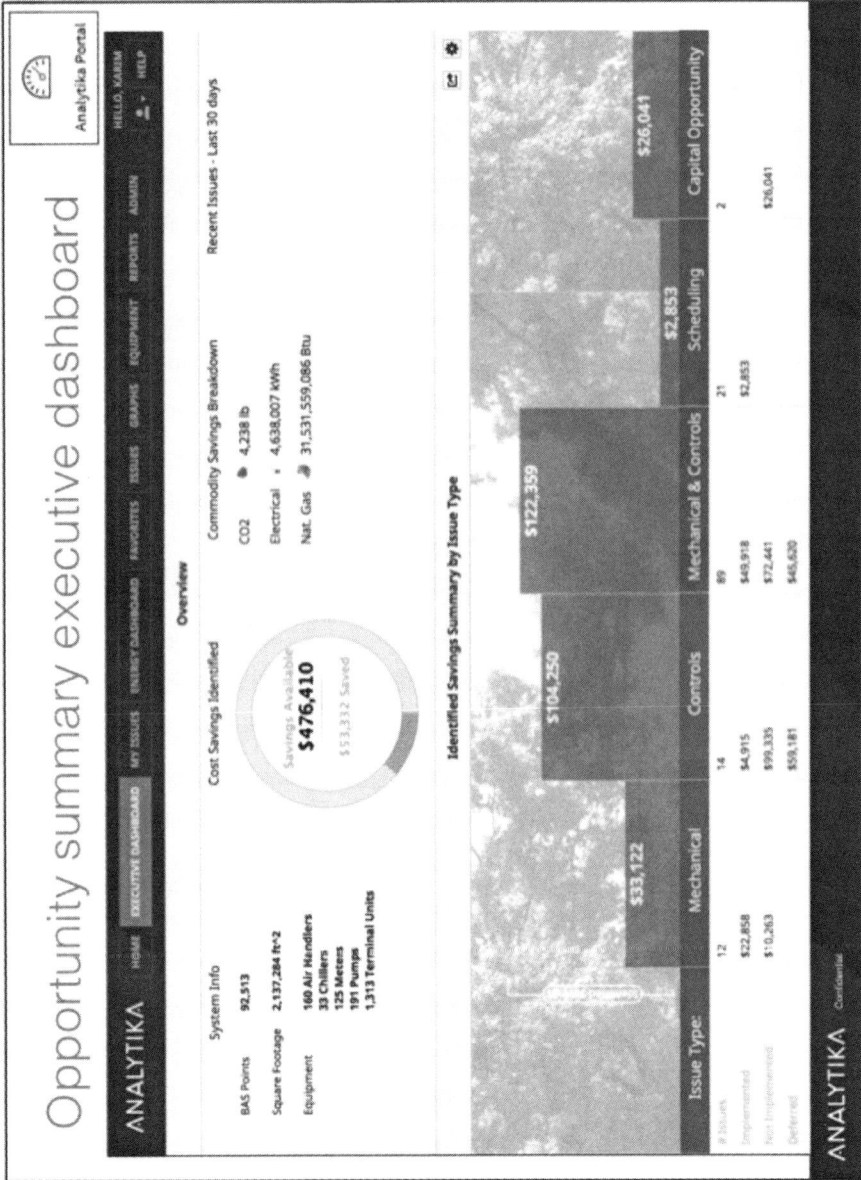

Figure 22-3

To illustrate these concepts in a real world example, consider the case study information provided below.

**Medical Center Saves $1.3 Million in**
**Annual Energy Costs While Improving Patient Comfort**
This medical center (MC)* is a 500-bed academic medical center providing pediatrics, primary and advanced medical care. MC is a large provider of trauma care and is affiliated with a prestigious university. The medical center has almost 1 million outpatient visits per year and the medical center campus includes over 2.5 million square feet located in the northeastern United States.

THE CHALLENGE

According to the EPA: "Health care organizations spend over $6.5 billion on energy each year to meet patient needs. Every dollar that a nonprofit health care organization saves on energy is equivalent to generating new revenues of $20 for hospitals or $10 for medical offices."

MC's guiding principles include efficiency and sustainability through a goal to reduce $CO_2$ emissions by 20% by 2020. To this end, MC was interested in implementing targeted energy conservation measures and technologies, which help achieve the sustainability goals while taking into account MC's specific needs.

MC also committed to the Healthier Hospitals Leaner Energy Challenge. The Healthier Hospitals Initiative (HHI) is "a national campaign to implement a completely new approach to improving environmental health and sustainability in the health care sector."

As a result of these goals and efforts, here is a summary of MC's achievements:

- Financial summary
  — Total energy savings: $1.3 million (annual)
  — Simple payback: 0.3 years
  — Net present value: $1,221,237

- Operational benefits
  — Sustainability and environmental stewardship: Achieved

---

*Due to confidentiality agreements, we are not able to provide the client name. We will refer to the organization as MC throughout this case study.

3,400+ metric tons in annual $CO_2$ emissions reduction, which is the equivalent of taking 715 cars off the road.

— Utility metering: Provided campus wide metering data by building and submeters which assisted fault detection and internal billing.

— Reporting: Assisted in joint commission reporting by providing analysis of specific zone conditions, including airflows, relative humidity levels, and temperatures for the operating rooms.

Cimetrics was selected to provide its Analytika Pro solution for 5 buildings comprising over 800,000 square feet including inpatient and outpatient care, medical offices, and the central plant. Cimetrics collaborated with MC, JCI and Siemens, their building automation system providers, to connect to and collect sensor and actuator data from over 9,400 physical points. Data were collected every 15 minutes, 24 hours a day, 365 days a year, totaling more than 900,000 data samples per day. The following systems were monitored: 41 air handling units, 19 hot water pumps, 11 chillers, 35 chilled water/condenser water pumps, 10 heat exchangers, 125 exhaust fans, 1,000 terminal units, 116 fume hoods, and other miscellaneous equipment. In addition, 2 chilled water meters, 1 condensate return meter, 52 electric meters, 17 steam meters, and 24 water meters were monitored.

Over 1,000 analytika software algorithms then analyzed the data to identify opportunities to reduce energy consumption, improve environmental conditions and reduce operations and maintenance costs. Analytika also uncovered potential equipment problems, quantified improved patient comfort, improved operating room zone conditions, and provided opportunities for profitable retrofit projects.

Experienced Cimetrics engineers leveraged analytika software to identify opportunities, determine root cause, and calculate annual savings impact. Actionable recommendations were documented and provided to the client both through online and offline channels. Cimetrics' role didn't end with providing recommendations; Cimetrics engineers engaged with the client team on a regular basis to help answer questions, coordinate implementation, and provide regular feedback on progress

Here is an example of an individual analytika issue—excess hours of operation for air handling units:

Four large air handling units (AHUs) serve an outpatient building. All were operating 24/7. However, there was no patient care at night and minimal building usage on weekends. The AHUs were left on 24/7 to serve magnetic resonance imaging (MRI) equipment and other high heat gain loads in the building. This issue was not detected because there were no temperature complaints; however, it was identified with analytika.

MRI machines and other hospital equipment can create substantial, unwarranted energy consumption if HVAC equipment is not specifically designed to exclusively serve that equipment.

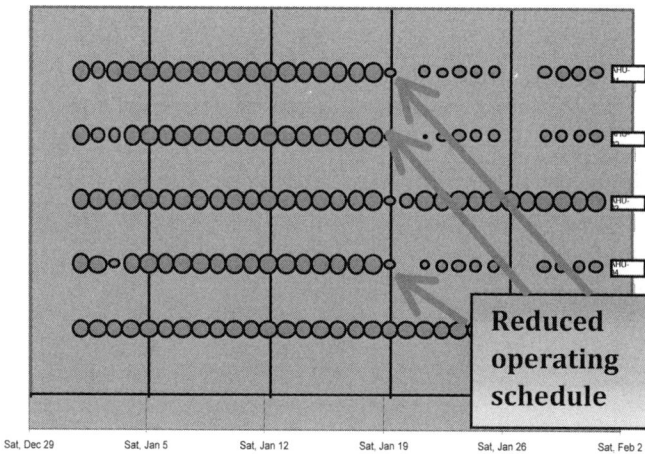

**Figure 22-4**

SOLUTION

Cimetrics worked with MC staff and the controls vendor to change the AHU sequence of operation to limit the equipment run times to only occupied hours and allow for override of the AHUs to maintain zones within the required temperature and relative humidity range.

Annual energy savings achieved: **$261,278**
Annual carbon emissions reduction: **605 metric tons**

# Index